Teaching STEM Education through Dialogue and Transformative Learning

This book focuses on reconceptualising the teaching of STEM education through dialogue and transformative learning, presenting examples of research from Mexico and the UK.

It centres on research which introduces critical pedagogies in the teaching of STEM, where in the past there has been an over-emphasis on content and a technicist perspective on science. The research in this book considers critical and dialogic approaches to teacher education for STEM subjects and emphasises the crucial role that teachers play in improving life chances for marginalised young people and their communities. STEM education is not just a way of improving a country's GDP, but if taught through dialogic and transformative pedagogies it can enable teachers to empower students to improve their own lives.

The collaboration between these two countries is timely and comes as Mexico is developing and emerging as a key global economic nation. The work presented here engages in theoretical and empirical work that has application beyond the two countries. This book was originally published as a special issue of the *Journal of Education for Teaching*.

Catherine Montgomery is Professor of International Higher Education in the Department of Education at the University of Bath, UK. She has worked on international and comparative research projects in Denmark, Mexico and Vietnam including the British Council project which underpins the research for this book, focusing on the role of dialogic STEM education in addressing social and cultural disadvantage in Mexico and the UK.

Juan Manuel Fernández-Cárdenas is Full Professor of Education at Tecnológico de Monterrey, Mexico. He has led several research projects and has published more than 75 research articles, refereed chapters and books. His research interests include the conformation of communities of practice mediated by the use of digital technology, in particular in STEM education.

Teaching STEM Education through Dialogue and Transformative Learning

Global Significance and Local Interactions in Mexico and the UK

Edited by
Catherine Montgomery and Juan Manuel Fernández-Cárdenas

Routledge
Taylor & Francis Group
LONDON AND NEW YORK

First published 2019
by Routledge
2 Park Square, Milton Park, Abingdon, Oxon, OX14 4RN

and by Routledge
52 Vanderbilt Avenue, New York, NY 10017

Routledge is an imprint of the Taylor & Francis Group, an informa business

British Library Cataloguing in Publication Data
A catalogue record for this book is available from the British Library

ISBN13: 978-0-367-26356-0

Typeset in Myriad Pro
by Newgen Publishing UK

Publisher's Note
The publisher accepts responsibility for any inconsistencies that may have arisen during the conversion of this book from journal articles to book chapters, namely the inclusion of journal terminology.

Contents

Citation Information

The chapters in this book were originally published in the *Journal of Education for Teaching*, volume 44, issue 1 (February 2018). When citing this material, please use the original page numbering for each article, as follows:

Editorial

Peter Gilroy
Journal of Education for Teaching, volume 44, issue 1 (February 2018) p. 1

Introduction

Teaching STEM education through dialogue and transformative learning: global significance and local interactions in Mexico and the UK
Catherine Montgomery and Juan Manuel Fernández-Cárdenas (England and Mexico)
Journal of Education for Teaching, volume 44, issue 1 (February 2018) pp. 2–13

Chapter 1

The social construction of a Teacher Support Team: an experience of university lecturers' professional development in STEM
Elvia Castro-Félix and Harry Daniels (Mexico and England)
Journal of Education for Teaching, volume 44, issue 1 (February 2018) pp. 14–26

Chapter 2

Developing a material-dialogic approach to pedagogy to guide science teacher education
Lindsay Hetherington and Rupert Wegerif (England)
Journal of Education for Teaching, volume 44, issue 1 (February 2018) pp. 27–43

Chapter 3

Creating a dialogic environment for transformative science teaching practices: towards an inclusive education for science
Cristina G. Reynaga-Peña, Marisol Sandoval-Ríos, José Torres-Frías, Carolina López-Suero, Adrián Lozano Garza, Maribel Dessens Félix, Marcelino González Maitland and Jorge G. Ibanez (Mexico)
Journal of Education for Teaching, volume 44, issue 1 (February 2018) pp. 44–57

Chapter 4

STEM outreach activities: an approach to teachers' professional development
Farzana Aslam, Arinola Adefila and Yamuna Bagiya (England)
Journal of Education for Teaching, volume 44, issue 1 (February 2018) pp. 58–70

Chapter 5

The role of non-formal contexts in teacher education for STEM: the case of horno^3 science and technology interactive centre
Claudia Fernández-Limón, Juan Manuel Fernández-Cárdenas and Alma Adrianna Gómez Galindo (Mexico)
Journal of Education for Teaching, volume 44, issue 1 (February 2018) pp. 71–89

Chapter 6

Public dialogue with science and development for teachers of STEM: linking public dialogue with pedagogic praxis
Richard Watermeyer and Catherine Montgomery (England)
Journal of Education for Teaching, volume 44, issue 1 (February 2018) pp. 90–106

Chapter 7

Innovating science teaching with a transformative learning model
Sandra Gudiño Paredes (Mexico)
Journal of Education for Teaching, volume 44, issue 1 (February 2018) pp. 107–111

Chapter 8

A hybrid and flipped version of an introductory mathematics course for higher education
N. Patricia Salinas Martínez and Eliud Quintero Rodríguez (Mexico)
Journal of Education for Teaching, volume 44, issue 1 (February 2018) pp. 112–117

Chapter 9

Developing non-formal education competences as a complement of formal education for STEM lecturers
Roy Alonso Terrazas-Marín (Mexico)
Journal of Education for Teaching, volume 44, issue 1 (February 2018) pp. 118–123

Notes on Contributors

Arinola Adefila is a researcher at the Centre for Global Learning: Education and Attainment at Coventry University, UK.

Farzana Aslam is Associate Head of School-Student Experience in the School of Computing, Electronics and Mathematics at Coventry University, UK.

Yamuna Bagiya is a mathematics and statistics assistant in the School of Computing, Electronics and Mathematics at Coventry University, UK.

Elvia Castro-Félix is affiliated with the Department of Engineering and Science at Monterrey Institute of Technology and Higher Education at San Luis Potosi, Mexico.

Harry Daniels is Professor of Education in the Department of Education at the University of Oxford, UK.

Maribel Dessens Félix is affiliated with the Centro Regional de Formacion Profesional Docente de Sonora, Mexico.

Juan Manuel Fernández-Cárdenas is Full Professor of Education at Tecnológico de Monterrey, Mexico.

Claudia Fernández-Limón is Educational Services Manager and a researcher at the Museo del Acero horno^3, Mexico.

Peter Gilroy is Visiting Professor at the Universities of Roehampton, Hull and Plymouth, UK, and editor of the *Journal of Education for Teaching*.

Alma Adrianna Gómez Galindo is a researcher at the Center for Research and Advanced Studies of the National Polytechnic Institute at the Cinvestav Unidad Monterrey, Mexico.

Marcelino González Maitland is affiliated with the Centro Regional de Formacion Profesional Docente de Sonora, Mexico.

Sandra Gudiño Paredes is a research professor in the School of Humanities and Education at Tecnologico de Monterrey, Mexico.

Lindsay Hetherington is a senior lecturer in Science Education at the Graduate School of Education at the University of Exeter, UK.

Jorge G. Ibanez is a professor in the Departamento de Ingenieria Quimica, Industrial y de Alimentos at the Universidad Iberoamericana Campus Ciudad de Mexico, Mexico.

Carolina López-Suero is a professor in the Departamento de Ingenieria Quimica, Industrial y de Alimentos at the Universidad Iberoamericana Campus Ciudad de Mexico, Mexico.

Adrián Lozano Garza is affiliated with the Department of Escuela de Humanidades y Educacion at Tecnologico de Monterrey, Mexico.

Catherine Montgomery is Professor of International Higher Education in the Department of Education at the University of Bath, UK.

Cristina G. Reynaga-Peña is a researcher in the Department of Escuela de Humanidades y Educacion at Tecnologico de Monterrey, Mexico, and CINVESTAV Unidad Monterrey, Mexico.

Eliud Quintero Rodríguez is a curriculum development specialist in the Department of Mathematics at Tecnologico de Monterrey, Mexico.

N. Patricia Salinas Martínez is a professor and educational researcher in the Department of Mathematics at Tecnologico de Monterrey, Mexico.

Marisol Sandoval-Ríos is a research professor at the Centro Regional de Formacion Profesional Docente de Sonora, Mexico.

Roy Alonso Terrazas-Marín is a researcher in the School of Information Technologies at the Technologic University of Chihuahua, Mexico.

José Torres-Frías is a professor in the Departamento de Cultura, Justicia y Democracia at the Centro Universitario del Norte at the Universidad de Guadalajara, Mexico.

Richard Watermeyer is a reader in the Department of Education at the University of Bath, UK.

Rupert Wegerif is Professor of Education in the Faculty of Education at the University of Cambridge, UK.

Preface

In this Special Issue we welcome as Guest Editors Catherine Montgomery and Juan Manuel Fernández-Cárdenas. Considerable additional resources have been poured into developing Science, Technology, Engineering and Maths (STEM) education which, by implication, suggests that these additional resources, especially in teacher education, may well have been diverted from other forms of education that have not been so privileged and might indeed have been marginalised. Such marginalisation would reinforce C. P. Snow's notion (1959) of the two separate and distinct cultures of science and the humanities, and especially concerning if policy makers are encouraging teacher educators to develop just one element of a state's overall culture.

Their collection of papers addresses this important issue by drawing on a British Council research project which used the STEM context to link the separate concepts of educational dialogue and transformational learning. In doing so the various authors have focussed on how teacher education in Mexico and the UK are developing STEM education through these two concepts and, most importantly, how these concepts might help in addressing those marginalised by an over-emphasis on STEM education.

Reference

Snow, C. P. 1959. *The Two Cultures*. London: Cambridge University Press.

Peter Gilroy

Introduction

Teaching STEM education through dialogue and transformative learning: global significance and local interactions in Mexico and the UK

Catherine Montgomery and Juan Manuel Fernández-Cárdenas
(England and Mexico)

Science, Technology, Engineering and Maths (STEM) education has attracted increasing attention in the last two decades, partly because the development of STEM itself, particularly in terms of Science and Technology, is often associated with national economic growth and human skills development. There is an assumed, generic link between STEM and the fostering of productivity, technological innovation and the growth of national economies and their GDP (Freeman, Marginson, and Tytler 2015). STEM education is thus becoming an increasingly significant focus for governments and for educational policy worldwide, with a high proportion of educational research and practice funding being earmarked to develop STEM. Despite the predominance of the claim that STEM education is integral to economic futures, there are critical perspectives on this economic argument in support of STEM education (Roschelle et al. 2011). Some research evidence reveals that overall educational quality is a more potent influence on economic outcomes (Hanushek and Woessmann 2012) and that innovative curricula and pedagogies are what is needed (Marginson et al. 2013). A number of large funded research projects over the last decade such as epiSTEMe in England (Cambridge University) have aimed to assess the impact of pedagogic change in STEM subjects or to intervene in the delivery of STEM as in the case of the STEM for Diversity project (see Hetherington and Wegerif, 2018). Other researchers claim that in many contexts education is not the predominant determining factor in economic growth and the impact of education on the economy is bounded by scope and time (Bevan 2011; Brown, Lauder, and Ashton 2012).

It is against this background that STEM education has become a piece in the complex jigsaw of globalisation. Globalisation itself is variously cast as a panacea for inequality (Zeidler 2016) and is also seen to be propagating as many inequalities as it is solving (Marginson et al. 2013; Stromquist and Monkman 2000). STEM education is similarly contradictory, being seen as a universal driving force for international development, but also remaining at the complex nexus of social, political and cultural inequalities with issues of gender and the education of indigenous populations being of particular concern. A recent UNESCO announcement noted the centrality of science education in the global sustainable development agenda, with the head of the United Nations' educational and scientific agency emphasising the importance of science centres and museums in developing skills and capacities, particularly for girls and marginalised communities (Alexovich 2016). Despite the evident significance of STEM, women and girls continue to be under-represented in fast-growing STEM jobs and with the increase in 'disruptive technologies', including robots and artificial intelligence, it will be marginalised groups that will be most affected (Amerasinghe 2016). Not only do women, diverse ethnicities and other marginalised groups continue to be under-represented in STEM sectors, but their concentration in low skilled employment is a vicious circle in the reproduction of social inequalities and disadvantage (Kersley and Shaheen 2014). Expanding engagement and emphasis on STEM education will be critical if there are to be improvements in the quality of work for under-represented groups and communities.

This special issue focuses on reconceptualising the teaching of STEM education through dialogue and transformative learning, presenting examples of research from Mexico and the UK. It centres on research which introduces a critical approach to pedagogies in the teaching of STEM, where in the past there has been an over emphasis on content and a more disciplinary perspective on science. It is important to note that what is meant or understood by the concept of STEM itself is not always clear; for some, STEM is synonymous with Science and for others it is interchangeable with Technology (Akgun 2013). The term STEM is used in many different ways and the papers in this special issue underline the fact that STEM, both as a concept and as an educational issue is strongly embedded in its social, cultural and political context (Khishfe et al. 2017).

The research in this special issue considers critical and dialogic approaches to teacher education for STEM subjects in Mexico and the UK and emphasises the crucial role that teachers play in improving life chances for marginalised young people and their communities. STEM education may be held up as a means of improving a country's GDP but if taught through dialogic and transformative pedagogies it can enable teachers to empower students to improve their own lives.

This special issue draws on research from a British Council funded project which linked the conceptual areas of educational dialogue and transformative learning in the context of STEM education in Mexico and the UK (Fernández-Cárdenas and Montgomery 2015, 2016). There has been considerable research work developed separately in educational dialogue and in transformative learning, but these two areas have not been brought together in the context of the study of STEM subjects in teacher education nor have these linked ideas been applied in Mexico and the UK. The collaboration between these two countries is timely and comes at a time when Mexico is developing and emerging as a key global economic and political nation. The special issue compiles a collection of articles focusing on educating teachers of STEM in an international comparative perspective, reflecting on the current state of teacher education in STEM subjects in Mexico and the UK and examining the sorts of innovations and new approaches that are being introduced in the two countries, particularly in dialogic teaching and transformative learning. There is also an emphasis here on how dialogic and transformative approaches in STEM education can improve opportunities for groups excluded or marginalised in STEM education through the refocusing of teacher education for STEM. This introductory paper aims to set the scene for the papers which follow by considering the policy context of teacher education in both Mexico and the UK and analysing the key themes that will be drawn out in this issue.

STEM and teacher education policy in England and Mexico

As discussed above, the grand narratives outlining the benefits of STEM and STEM education in improving national, industrial and community prosperity are prevalent but this discourse is not accompanied by a strong discourse in policy for education. In the UK, there appears to be a lack of systematic actions in relation to the policy on how teachers should be educated to engage with the fast-moving world of STEM. According to a study carried out by the Council for Science and Technology (2000)

> there was a concern among science teachers about how they could develop personally and professionally throughout their careers. Teachers relied on local networks of informal contacts, either in-school or between schools, and a number of school-based training days which, because of their whole-school nature, rarely dealt with subject-specific issues. (Tomei, Dillon, and Dawson 2014, 173; see also Dillon et al. 2000)

This indicates that teachers feel a lack of guidance and direction from policy for STEM education and some of the papers in this special issue explore both the effects of this (Aslam, Adefila, and Bagiya, 2018) and some possible innovative approaches to addressing this (Fernández Limón, Fernández-Cárdenas, and Gómez Galindo, 2018; Watermeyer and Montgomery, 2018). In the current education policy in the UK, reference to STEM education is located in the Department for Business, Energy and Industrial Strategy (BEIS), the department that recently replaced the UK Department for Business Innovation and Skills. A policy paper was published in 2012 and updated in May 2015 entitled 'Public understanding of Science and Engineering' which states:

> Science and research are major contributors to the prosperity of the UK. For our prosperity to continue, the government believes we need high levels of skills in science, technology, engineering and maths (STEM), and citizens that value them. (Department for Business, Innovation and Skills 2015)

In terms of the ways that this policy is enacted, the policy paper notes that the approach to promoting STEM will be to engage the public in science and engineering through the British Science Festival and the National Science and Engineering Week and events that promote science and raise the public's awareness of science issues (see Watermeyer and Montgomery, 2018). The approach also involves funding of 4 independent national academies which are the Royal Society; the British Academy; the Royal Academy of Engineering and the Academy of Medical Sciences. The encouragement of science in schools and the funding of programmes and events that inspire students to study STEM subjects is also mentioned as a key priority. Despite this foregrounding of the significance of STEM and engaging with education, there is a notable absence of links with the Department of Education or any links with teacher education policy. An initial analysis of the policies of the UK Department of Education published since 2011 show that the strategy has been segmented into regions and it was only possible to locate specific recommendations from Northern Ireland. These recommendations are centred on (a) coordinating business links so that the demand for STEM graduates increases, (b) managing STEM sector attractiveness by working with museums and schools in innovative ways, and (c) facilitating CPD development, contextualising educational activities to meet the demands of communities (Department of Education 2011). Given this segmented strategy in STEM education policy in the UK, it is easy to see how challenges in translating this policy into teacher education for STEM could be generated. As far as policy on teacher education in England is concerned, the most recent policy guidance on teacher education was the Carter Review which appeared in July 2016, giving the framework for core content for Initial Teacher Training (ITT) but there is a notable absence here on guidance for specific subjects such as STEM (Carter 2015). Furthermore, there is no guidance in this policy on the enactment of subject specific pedagogy for any particular disciplines or subjects resulting in a silence on more complex interdisciplinary pedagogy for STEM education and its implications for teacher education. In teacher education programmes in England, guidance on STEM is compartmentalised into single disciplines such as Science and Maths, leaving an interdisciplinary approach to remain as an ideal. In 2013 the UK government moved towards more focus on school-based teacher training and this complicates the context considerably, meaning that the role that Higher Education could play in the debate around STEM education and its implications for teacher education is also increasingly limited. The segmented policy on STEM for teacher education is thus even more problematic and results in greater fragmentation of approaches and understandings.

Notwithstanding the apparent gaps in education policy and teacher education policy in England, this does not mean that there are no initiatives around public engagement with STEM and STEM education. There are extensive STEM networks and STEM activities in communities, museums and informal science centres across the UK which are aiming to encourage engagement with STEM. The organisation STEM Ambassadors ('STEM Ambassadors STEM' 2017) is an example of such an organisation and is a government funded group which organises events, an ambassadors' scheme, funds after school clubs and offers resources and CPD for teachers. The UK Department for Education and the Wellcome Trust support the National STEM Centre which runs a programme of CPD activities and has a strong resource collection in its e-library. Part of this collection relates to policy for teacher education ('STEM Education: National Policies and Reports' 2010) but much of this is out of date. The National STEM Learning Network ('Impact of the National STEM Learning Network' 2017) also provides a wide range of activities for students and teachers but it is not clear what policy priorities underpin these activities. Thus, despite the funding being given to these sorts of activities which demonstrate the UK government's commitment to supporting the development of STEM education, there are tenuous links between these activities and teacher education. There may be informal links through schools outwards to these initiatives but in terms of educating teachers around how to engage well with these resources, there is a gap (see Aslam, Adefila, and Bagiya, 2018).

An international comparison of the situation with policy on teacher education for STEM shows that there are similarities and differences in Mexico. Historically, science education in primary and secondary schools in Mexico has not been seen as a priority for national development. Spanish and Mathematics, as disciplinary fields, without an applied multidisciplinary approach (as is the case of STEM education), have been considered the most important subjects for developing competencies for successful careers and for solving problems in everyday life. Thus, little time is allowed for the teaching of science in basic education classrooms and CPD efforts have not focused sufficiently on realising the importance of science and STEM education for an economy based on applied research and innovation (Flores-Camacho 2012).

More recently, a new educational policy has been designed in Mexico (Secretaría de Educación Pública 2017) and this highlights the speed of the social and academic changes which have become part of our everyday lives in a globalised world. The policy stipulates that teachers should have the competencies to be able to solve problems derived from societal change, and to use these processes as exemplars to construct knowledge with their students. Teachers need to be capable of teaching complex curricular subjects to students and should also be able to promote their active engagement in the solution of the problems of their communities. In order to achieve these goals, teachers participate in a Continuous Professional Development scheme called 'Teacher's Professional Service' ('Servicio Profesional Docente'), which is aimed at the development of competencies through the evaluation and the modelling of innovative educational practices through two different schemes. The first scheme draws on the offer of courses available at teacher training centres (Centro de Capacitación y Actualización de Maestros, CECAM) where a catalogue of CPD courses is offered for teachers in order to promote a situated, self-regulated learning, through collaboration and social participation. The second scheme takes place directly in schools, and is carried out through the cultivation of communities of practice where teachers engage in reflective experiences about their everyday life at school and the development of pedagogical practices (Posner 2004). Reflective practice is discussed as part of this CPD community and led by the head teacher in each educational setting. In general, the pedagogy which the Mexican government hope to implement as part of this new model draws on the use of real world problems to be addressed as part of the curriculum, relating disciplinary knowledge with the construction of situated solutions. However, these goals are far from what actually occurs in STEM education in Mexico; and policy is similarly unclear on teacher education with only very general statements presented to guide the educational Mexican system and an acknowledgment that it is difficult to keep pace with the speed of change, thus:

> the accelerated development of science and technology have had a profound impact in everyday life of the inhabitants of all countries on earth. This area of human knowledge has developed with great speed in contrast with what used to happen in previous periods with a more gradual rhythm of change. (Secretaría de Educación Pública 2017, 129)

The most explicit policy relating to STEM and teacher education for STEM in Mexico is on technology education. Teachers are expected to master the use of ICT, and to incorporate digital open resources into their lessons. Also, training in robotics is being promoted through an initiative called '@prende 2.0', which involves the equipment of 3000 special ICT rooms in selected schools, with high speed internet, specialised robotics pieces organised in collections, and computational languages for programming solutions and building apps. This is a very recent initiative that has not been assessed yet, but resembles the goals of an emblematic educational programme called 'Enciclomedia', based on the massive set up of interactive whiteboards (IWBs) in schools across the nation. IWBs added a value to the educational process, but in a limited fashion determined by the pedagogical background and knowledge of each teacher (Fernández-Cárdenas and Silveyra-De La Garza 2010). Despite this, there is no single explicit reference to STEM in the policy documents produced by Secretary of Education in Mexico.

At a societal level, and as in the UK, several non-governmental organisations are working in favour of advancing the agenda on STEM Education in Mexico. For instance, Innovec ('Innovec' 2017) is a civil association aiming to advance research, innovation, and the development of strategies to improve the teaching of scientific activities in basic education for children and teenagers. It was originally funded by the Mexico-United States Association for Science and since 2002 it has been delivering CPD programmes and workshops for pupils and teachers, as well as designing assessment practices looking at science as an inquiry-based endeavour which requires links with communities and academic bodies in universities and research institutes.

Similarly, 'Programa Adopte un Talento' ('PAUTA' 2017) is an association of scientists, pedagogues, educational psychologists, and teachers, who are devoted to offering seminars and workshops for pupils and teachers with the objective of developing STEM competencies. Pauta is based in 5 states in Mexico and is working as a non-profit organisation in underdeveloped settings, promoting socioeconomic development through the appropriation of scientific and technologic tools for solving problems emerging from the community and in close alliance with local schools.

Finally, as a third example of Mexican initiatives, 'Sophie' ('Proyecto Pedagógico Sophie UNAM, Ciencia Para Niños y Jóvenes' 2017) is a project which aims to introduce children and teenagers to science in the disciplines of Physics, Chemistry, Mathematics, and Biology, as well as promoting the development of intelligence, creativity, and critical thinking. Sophie is a UNESCO project led by the National Autonomous University of Mexico and uses experimentation and play to enhance curiosity and exploration, helping students to focus their attention in inquiry based activities.

All these initiatives have laudable missions in relation to helping young people and teachers to develop competencies in STEM subjects. However, it is problematic that these organisations are not linked to and do not receive sustainable support from the government, and they are also not articulated in government policy on teacher education for STEM. If STEM is going to make a societal change, it is necessary to increase the scale of the educational impact, growing from a selection of interesting cases and associations towards a wider ranging systemic effect.

Challenges of moving forward with teacher education for STEM

As indicated by the discussion above there are direct challenges posed by the structures and formats of teacher education in the area of STEM. A major challenge in STEM education in both Mexico and the UK is the quality of teachers, their engagement with innovative pedagogies for STEM and teachers' own individual understandings and expertise in understandings of Science (see Aslam, Adefila, and Bagiya, 2018; Fernández Limón, Fernández-Cárdenas, and Gómez Galindo, 2018; Watermeyer and Montgomery, 2018). Professionals with in-depth expertise in STEM subjects are tempted to follow careers in STEM itself and the status of the teaching profession in both countries discourages 'real scientists' from pursuing a career in teaching. In particular, away from the large cities, in rural settings in Mexico teachers themselves are marginalised and it is striking to see that the majority of teachers come from lower socio-economic backgrounds and are not closely engaged or linked to communities of scientific education, and therefore have problems in catching up with dominant discourses of science (INEE 2015; Torres del Castillo 1998). In the UK, there are teacher shortages in most STEM subjects and even at 'A Level' (16–18 years) teachers are required to teach in disciplines that are not their own. In both Mexico and the UK, teacher education is perceived as a second-class option and society tends to see teachers as second-class professionals, especially in contrast to 'real scientists' (see Watermeyer and Montgomery, 2018). Papers in this special issue suggest that the weak links between teacher education, science and the knowledge base of teachers are a crucial element in the continuing inequalities around STEM education (see Fernández Limón, Fernández-Cárdenas, and Gómez Galindo, 2018; Watermeyer and Montgomery, 2018).

Given the difficulties in addressing the complex inequalities and marginalisation associated with the development of STEM, it appears that there are challenges in moving forward with exploiting the potential of the current interest and investment in STEM education. Much focus in regional economic development strategy aims to grow the scientific, engineering and manufacturing sectors rather than improving approaches to education. Alongside this there has been some significant infrastructure investment around sector development and 14–19 education and training but this has occurred in regional pockets in both the UK and Mexico. Emerging research from organisations (Pike et al. 2016) suggests more work needs to be done to ensure that investments benefit the most disadvantaged elements of local communities.

Dialogic and communitarian STEM education

This special issue aims to construct STEM education as dialogue embedded in both formal and non-formal communities and built on constructive partnerships between teacher educators, alternative non-formal science education settings in the community and the more formal bastions of scientific research and knowledge in the university. In this model of communitarian STEM education, particular significance is placed on dialogic understandings, both in the context of community relations, but also in terms of the pedagogies employed in STEM in schools and universities. Bakhtinian dialogism encourages the analysis of patterns of teacher-student communication in naturalistic classroom settings (Skidmore and Murakami 2016; Hetherington and Wegerif, 2018) but this special issue calls for a broader understanding of dialogism that extends to interaction between teacher educators, teachers, community actors and more formal domains of knowledge in the university and society, promoting a Freirean style of communitarian dialogue as emancipation. Freire (1989) proposed that education should develop a critical consciousness in which teachers empower students to surface their personal understandings of their own learning experiences but this approach is rarely applied in STEM education. Dialogic teaching moves away from the privileging of one voice (a monologic space) where the dominant voice is usually the teacher, towards the development of a dialogic space in teaching where students and teachers interact around the curriculum (Cowie and Van der Aalsvoort 2000; Hetherington and Wegerif, 2018). In addition, studies in education following a dialogic perspective have highlighted the importance of language positioning in the context of the teaching relationship, as well as the ethics involved in a dialogic space in which two or more voices can express themselves without trying to silence each other (Bakhtin 1993, 2010; Castro Félix and Daniels, 2018; Reynaga-Peña et al., 2018).

These dialogic transformative partnerships aim to move away from the conception of STEM as content towards the relocation of STEM in local settings. A crucial element of this is enabling teachers and teacher educators to understand their own role and the role of their own knowledge in STEM education, not as conveyors of knowledge but as dialogic partners in co-construction in the local community.

This special issue

This special issue aims to build an understanding of how dialogic approaches can be embedded in teacher education for STEM and how this can be more strongly embedded in local communities, engaging teachers with structures both within and outside of the classroom and school in order to develop a locally relevant dialogic pedagogy for STEM which could be more effective in addressing marginalisation.

The special issue opens with focus on teacher education for STEM in formal contexts, with a higher educational perspective on how communities of engineering and science teacher educators in university settings can transform their understandings of their own practice. Castro-Félix and Daniels offer a Vygotskian, socio-cultural account of how teacher support teams can improve university tutors' use of joint activities in order to make changes in their teaching and thus improve student performance. This paper shows how it is possible to use the affordances of dialogue within the boundaries of formal higher educational settings, constructing common goals that could bring about continuous improvement in teaching practices in STEM disciplines.

In the second paper in this collection we continue to consider teacher education for STEM in formal educational settings with a close-up view on teachers working with dialogic STEM in the classroom. Hetherington and Wegerif argue that science teacher education and professional development should pay more attention to the material-dialogic relationships in the learning that emerges in science classrooms. The paper suggests that it is crucial for trainee teachers to become competent in managing the materiality of classroom practice and that dialogic pedagogy could remain a theoretical concern if teachers fail to broaden their understanding of the complexities of the material in science. Hetherington and Wegerif's drawing together of a Bakhtinian dialogic pedagogy (Wegerif 2012) with Barad's material-discursive epistemology is an innovative theoretical approach which could provide an important new way of constructing dialogic practices in STEM education.

In the third paper Reynaga-Peña et al. focus attention on the use of dialogue in inclusive practices, specifically centring on STEM education for visually impaired learners. The case study presented underlines the importance of teacher peer and self-reflection around inclusion in science education, transforming and sensitising teachers' approaches through the creation of adult learning environments grounded on the principles of dialogic learning. The paper concludes that it is possible to stimulate a gradual transformation of teaching practices through CPD courses promoting self-awareness and critical reflection, which result in a willingness to change towards solidarity and social action.

In the fourth paper the special issue begins to focus on the interface between formal and non-formal settings with Aslam, Adefila and Bagiya examining the ways in which STEM teachers in secondary education perceive the impact of STEM engagement on their own professional development. Aslam et al. argue for the crucial significance of outreach work in STEM where teachers become the interface between the formal context of the school and community-based STEM activity and initiatives. Whilst there are many structural challenges in the ways that outreach work is being brought into formal education, the paper argues that when teachers become the bridge to outreach communities for their learners, this also brings benefits to the learning of the teachers themselves.

In the fifth paper in the special issue, the focus moves beyond formal educational settings to the role of more informal contexts in the education of teachers of STEM. For Fernández Limón, Fernández-Cárdenas and Gómez Galindo, the post-industrial science museum is a vehicle for broadening the perspectives of teachers. This paper reflects the commitment to create new forms of engagement and participation in science museums and science centres underlining the growing recognition that non-formal learning can play a significant role in

STEM education and enable teachers of STEM to come to understand important contemporary debates about science and technology.

Finally, the sixth paper in this issue draws our attention to the societal context and the role of public science in the education of teacher of STEM. Watermeyer and Montgomery focus on the ways in which initiatives around public engagement with science can provide principles for dialogue in teacher education. The paper considers how the engagement of teachers in public science can address issues of teacher knowledge and the paper suggests ways in which principles of good dialogue drawn from models of public engagement with science can be transposed to the teacher training context.

The special issue closes with three practice papers (Gudiño Paredes, 2018; Salinas Martínez and Quintero Rodríguez, 2018; Terrazas-Marín, 2018), presenting practical examples from Mexico of some of the ways in which science education for teachers is being approached differently. These practice papers illustrate the importance for teacher education of the link between families and science education, the role of digital technologies in modelling science for teachers and the possibilities offered from outdoor education for transforming teachers' approaches to science education.

The papers in this special issue aim to enable the reader to rethink their perspectives on teacher education for STEM. Here the emphasis is on collaboration and communication and the papers indicate that non-formal learning for teachers is critical and arises in and through social interaction, interaction that enables people to mutually engage in cooperation and co-participation and to become familiar with and understand seemingly technical and complex issues and problems (Falk et al. 2004; Gee 1999; Rennie et al. 2003). Placing the agency for directing STEM within the community could provide a catalyst for promoting STEM careers and genuinely opening opportunities for marginalised groups such as girls, women, lower socio-economic groups and indigenous or ethnic communities. However, the special issue also underlines the need for formal structures which enable this to happen.

The differences in the way in which globalisation is manifesting itself in the current moment are highlighted by STEM education and illustrated by the case studies presented in this special issue. The examples suggest that by working creatively with local contexts through harnessing the agency of community and non-formal educational settings the claims for the benefits of STEM education may be justified. Partnerships between community settings, teacher educators and science contexts will be a crucial part of this. For these communitarian interactions to succeed it will require a much more coordinated approach between the community, schools, universities, public science and corporations and crucially with government policy on teacher education for STEM.

Acknowledgements

We would like to acknowledge the input of Dr Paul Denley from the University of Bath on teacher education policy in the UK and Dr Cristina Reynaga-Peña of Tecnológico de Monterrey on teacher education policy in Mexico.

Disclosure statement

No potential conflict of interest was reported by the authors.

ORCID

Catherine Montgomery http://orcid.org/0000-0002-5829-6137
Juan Manuel Fernández-Cárdenas http://orcid.org/0000-0003-2044-1658

References

Akgun, O. E. 2013. "Technology in STEM Project-based Learning." In *STEM Project-based Learning: An Integrated Science, Technology, Engineering, and Mathematics (STEM) Approach*. 2nd ed., edited by R. M. Capraro, M. M. Capraro, and James R. Morgan, 65–75. DIVS. Rotterdam: Sense Publishers.

Alexovich, A. 2016. "UNESCO: Science Museums Key for Sustainable Development." *United Nations Sustainable Development* (blog). Accessed November 10, 2016. http://www.un.org/sustainabledevelopment/blog/2016/11/unesco-science-museums-vitally-important-for-sustainable-development/

Amerasinghe, S. 2016. "Women's Jobs at Risk from Tech Disruption." Text. Jobs and Development. Accessed December 7, 2016. http://blogs.worldbank.org/jobs/women-s-jobs-risk-tech-disruption

Aslam, F., A. Adefila, and Y. Bagiya. 2018. "STEM Outreach Activities: An Approach to Teacher's Professional Development." *Journal of Education for Teaching* 44 (1): 58–70.

Bakhtin, M. M. 1993. *Problems of Dostoevsky's Poetics*. Edited by Caryl Emerson. Minneapolis: University of Minnesota Press.

Bakhtin, M. M. 2010. *Speech Genres and Other Late Essays*. Austin: University of Texas Press.

Bevan, B. 2011. "Examining Links between STEM Education and the Economy | Current Research." Accessed June 1, 2017. http://www.relatingresearchtopractice.org/article/124

Brown, P., H. Lauder, and D. Ashton. 2012. *The Global Auction: The Broken Promises of Education, Jobs, and Incomes*. Oxford: Oxford University Press.

Carter, A. 2015. "Carter Review of Initial Teacher Training (ITT)." Accessed June 1, 2017. http://dera.ioe.ac.uk/21832/7/Carter_Review_16012015_Redacted.pdf

Castro Félix, E. G., and H. Daniels. 2018. "The Social Construction of a Teacher Support Team. An Experience of University Tutors' Professional Development in STEM." *Journal of Education for Teaching* 44 (1): 14–26.

Council for Science and Technology. 2000. *Science Teachers: Supporting and Developing the Profession of Science Teaching in Primary and Secondary Schools*. London: Department of Trade and Industry.

Cowie, H., and G. Van der Aalsvoort. 2000. *Social Interaction in Learning and Instruction: The Meaning of Discourse for the Construction of Knowledge*. Bingley: Emerald Group Publishing Limited.

Department for Business, Innovation and Skills. 2015. "2010 to 2015 Government Policy: Public Understanding of Science and Engineering – GOV.UK." May 8, 2015. Accessed October 11, 2017. https://www.gov.uk/government/publications/2010-to-2015-government-policy-public-understanding-of-science-and-engineering/2010-to-2015-government-policy-public-understanding-of-science-and-engineering

Department of Education. 2011. "Success through STEM. STEM Strategy in Response to the 'Report of the STEM Review'." Bangor: Department of Education – Northern Ireland.

Dillon, J., J. Osborne, R. Fairbrother, and L. Kurina. 2000. *A Study into the Professional Views and Needs of Science Teachers in Primary and Secondary Schools in England*. London: King's College London.

Falk, J. H., C. Scott, L. Dierking, L. Rennie, and M. Cohen Jones. 2004. "Interactives and Visitor Learning." *Curator: The Museum Journal* 47 (2): 171–198. https://doi.org/10.1111/j.2151-6952.2004.tb00116.x.

Fernández Limón, C., J. M. Fernández-Cárdenas, and A. A. Gómez Galindo. 2018. "The Role of Non-formal Contexts in Teacher Education for STEM: The Case of Horno3 Science and Technology Interactive Centre." *Journal of Education for Teaching* 44 (1): 71–89.

Fernández-Cárdenas, J. M., and C. Montgomery. 2015. "STEMk12dialogue: STEMk12dialogue." Accessed October 21, 2015. http://stemk12dialogue.blogspot.co.uk/p/description-of-workshop-this-workshop.html

Fernández-Cárdenas, J. M., and C. Montgomery. 2016. "STEMK12 Dialogue." Accessed October 17, 2017. http://www.stemk12dialogue.net/index.html

Fernández-Cárdenas, J. M., and M. L. Silveyra-De La Garza. 2010. "Disciplinary Knowledge and Gesturing in Communicative Events: A Comparative Study between Lessons using Interactive Whiteboards and Traditional Whiteboards in Mexican Schools." *Technology, Pedagogy and Education* 19 (2): 173–193.

Flores-Camacho, F., ed. 2012. *La Enseñanza de la Ciencia en la Educación Básica en México* [The Teaching of Science in Basic Education in Mexico]. México: Instituto Nacional de Evaluación Educativa.

Freeman, B., S. Marginson, and R. Tytler. 2015. *The Age of STEM: Educational Policy and Practice across the World in Science, Technology, Engineering and Mathematics*. London: Routledge.

Freire, P. 1989. "Paulo Freire's Speech given at the President's Forum at Claremont Graduate University at the Commencement Activities in Which He Was Awarded an Honorary Doctorate." Presented at the Honorary Doctorate for Paulo Freire, Claremont, CA, May.

Gee, J. P. 1999. *An Introduction to Discourse Analysis: Theory and Method*. London: Routledge.

Gudiño Paredes, S. 2018. "Innovating Science Teaching with a Transformative Learning Model." *Journal of Education for Teaching* 44 (1): 107–111.

Hanushek, E. A., and L. Woessmann. 2012. "The Economic Benefit of Educational Reform in the European Union." *CESifo Economic Studies* 58 (1): 73–109. doi:10.1093/cesifo/ifr032.

Hetherington, L., and R. B. Wegerif. 2018. "Developing a Material-dialogic Approach to Pedagogy to Guide Science Teacher Education." *Journal of Education for Teaching* 44 (1): 27–43.

"Impact of the National STEM Learning Network." 2017. Accessed June 1, 2017. https://www.stem.org.uk/news/impact-national-stem-learning-network

INEE. 2015. "Los Docentes En México. Informe 2015." [Teachers in Mexico. Report 2015]. Informes Temáticos. México: INEE.

"Innovec." 2017. Accessed July 12, 2017. http://innovec.org.mx/home/

Kersley, H., and F. Shaheen. 2014. *Addressing Economic Inequality at Root: 5 Goals for a Fairer UK*. London: New Economics Foundation/Friedrich Ebert Stiftung.

Khishfe, R., F. S. Alshaya, S. BouJaoude, N. Mansour, and K. I. Alrudiyan. 2017. "Students' Understandings of Nature of Science and Their Arguments in the Context of Four Socio-scientific Issues." *International Journal of Science Education* 39 (3): 299–334. doi:10.1080/09500693.2017.1280741.

Marginson, S., R. Tytler, B. Freeman, and K. Roberts. 2013. *STEM: Country Comparisons: International Comparisons of Science, Technology, Engineering and Mathematics (STEM) Education. Final Report*. Australian Council of Learned Academies.

"PAUTA." 2017. Accessed July 12, 2017. http://www.pauta.org.mx/

Pike, A., M. Coombes, T. Champion, D. Bradley, L. Robson, and C. Wymer. 2016. *Uneven Growth Tackling City Decline Findings*. York: Joseph Rowntree Foundation.

Posner, B. Z. 2004. "A Leadership Development Instrument for Students: Updated." *Journal of College Student Development* 45 (4): 443–456. doi:10.1353/csd.2004.0051.

"Proyecto Pedagógico Sophie UNAM, Ciencia Para Niños y Jóvenes." [Pedagogic Project Sophie UNAM, Science for Children and Youngsters]. Accessed July 12, 2017. https://sophie.unam.mx/

Rennie, L. J., E. Feher, L. D. Dierking, and J. H. Falk. 2003. "Toward an Agenda for Advancing Research on Science Learning in Out-of-school Settings." *Journal of Research in Science Teaching* 40 (2): 112–120. doi:10.1002/tea.10067.

Reynaga-Peña, C. G., M. Sandoval-Ríos, J. Torres-Frías, C. López-Suero, O. A. Lozano Garza, M. Dessens Felix, M. González Maitland, and J. G. Ibáñez. 2018. "Creating a Dialogic Environment for Transformative Science Teaching Practices towards Inclusive Education of Science." *Journal of Education for Teaching* 44 (1): 44–57.

Roschelle, J., M. Bakia, Y. Toyama, and C. Patton. 2011. "Eight Issues for Learning Scientists about Education and the Economy." *Journal of the Learning Sciences* 20 (1): 3–49. doi:10.1080/10508406.2011.528318.

Salinas Martínez, N. P., and E. Quintero Rodríguez. 2018. "A Hybrid and Flipped Version of a Preparatory Mathematics Course for Higher Education." *Journal of Education for Teaching* 44 (1): 112–117.

Secretaría de Educación Pública. 2017. "Modelo Educativo para la Educación Obligatoria." [Educational Model for Compulsory Education]. Comisión Nacional de Libros de Texto Gratuitos. Accessed July 12, 2017. https://www.gob.mx/cms/uploads/attachment/file/198738/Modelo_Educativo_para_la_Educacio_n_Obligatoria.pdf

Skidmore, D., and K. Murakami. 2016. *Dialogic Pedagogy: The Importance of Dialogue in Teaching and Learning*. Multilingual Matters.

"STEM Ambassadors, STEM." 2017. Accessed June 1, 2017. https://www.stem.org.uk/stem-ambassadors/

"STEM Education: National Policies and Reports, STEM." 2010. Accessed July 12, 2017. https://www.stem.org.uk/resources/collection/94082/stem-education-national-policies-and-reports

Stromquist, N. P., and K. Monkman. 2000. *Globalization and Education: Integration and Contestation across Cultures*. Lanham, MD: Rowman & Littlefield.

Terrazas-Marín, R. A. 2018. "Developing Non-Formal Education Competences as a Complement of Formal Education for STEM Lecturers." *Journal of Education for Teaching* 44 (1): 118–123.

Tomei, A., J. Dillon, and E. Dawson. 2014. "United Kingdom: An Example of the Impact of High Stakes Accountability Regimes on STEM Education." In *The Age of STEM: Educational Policy and Practice across the World in Science, Technology, Engineering and Mathematics*, edited by Brigid Freeman, Simon Marginson, and Russell Tytler, 161–177. London: Routledge.

Torres del Castillo, R. M. 1998. "Nuevo Papel Docente. ¿Qué modelo de formación y para qué modelo educativo?" [What Training Model and for What Educational Model?] *Perfiles Educativos* 82: 1–19.

Watermeyer, R., and C. Montgomery. 2018. "Public Dialogue with Science and Development for Teachers of STEM: Linking Public Dialogue with Pedagogic Praxis." *Journal of Education for Teaching* 44 (1): 90–106.

Wegerif, R. B. 2012. *Dialogic: Education for the Internet Age*. London: Routledge.

Zeidler, D. L. 2016. "STEM Education: A Deficit Framework for the Twenty First Century? A Sociocultural Socioscientific Response." *Cultural Studies of Science Education* 11 (1): 11–26. doi:10.1007/s11422-014-9578-z.

1 The social construction of a Teacher Support Team

An experience of university lecturers' professional development in STEM

Elvia Castro-Félix and Harry Daniels (Mexico and England)

ABSTRACT

This paper focuses on understanding and exploring how a group of university engineering and science tutor educators learn and assimilate new conceptions about their role in the face of the forces of globalisation that are transforming the system of higher education. This research paper adopts the notion of the Teacher Support Team (TST) as developed by Daniels and grounded in Vygotsky's sociocultural account of the social formation of mind. These structures of meaning provide insight into the role played by the context, the interactions, the needs and the demands of actual activities, agreements and learning processes that this group of STEM lecturers undertook as they sought to transform their usual teaching methods that were focused on individual and isolated work in order to create more innovative practices and impact on their students' performance. The analysis of this experience, which was based on the epistemological principles of the sociocultural approach, focused on the educational model that emerges from needs that are perceived and shared through the group's interactions, as well as the transitions that such a team undergoes in its actions and decisions.

Identifying the problem

Today's globalised world is ever-changing in many aspects. It is challenging to understand the professional capabilities required in different labour markets without considering the wider economic and social world and its calls for new socio-professional objectives. In our case, we focus on the transformation of educational systems and policies within a private university in western Mexico. This has had a profound effect on the new lecturers' professional role, with the introduction of a new strategy of curriculum formation which calls for enhanced knowledge exchange and collaboration. Such pressures on education make it possible to appreciate, among other issues, the reasons behind the emergence of new professional identities and social practices (Daniels and Garner 2013).

In the Mexican context, Science, Technology, Engineering and Math (STEM) university lecturers are being encouraged to focus on solving problems for society. They are expected to work collaboratively in groups in order to implement fundamental changes in their teaching practices. In this setting, there is considerable interest in examining and understanding how university lecturers learn and develop new conceptions of their practice in response to the demands of this globalised context. As a result, the concept of the *Teacher Support Team* (TST) developed by Daniels (1998) has been adopted and transformed to meet the needs of this local scenario. The TST approach carries with it a focus on context, social interaction, dialogue, roles taken on by lecturers, joint activities, obstacles and limitations that arise within relationship that seek, from the perspective of the lecturer, to make a change in teaching and improve student performance (Sannino and Ellis 2013).

The approach used in this study to examine the learning and cognitive development of lecturers participating in a 'team-learning' type of professional education seeks to avoid the traps of concentrating only on the individual or on mental processes in a vacuum (Moll 2013). The notion of a TST as proposed by Daniels, grounded as it is in Vygotsky's sociocultural theory, helps university staff to gain insights into ways in which deep changes in lecturers' understanding can be achieved through participation, activities and conversations with colleagues. The concept of social collaborative learning, which posits a process of development as part of the intellectual life of others with whom one interacts, represents another key foundation of this research. This paper aims to give an account of a professional educational model based on peer interaction and dialogue in academic work, its advantages and its impact on university lecturers' culture and practice.

Theoretical framework

The main focus of this research is on an intervention strategy that is configured through the interaction and dialogue between peers who are members of a university teaching staff in order to influence students' learning. The TST intervention was adopted in order to enhance the ways in which university lecturers learn and achieve new understandings of their function. The decision to use this theoretically driven construct was influenced by the advantages that it offers in understanding the context of the interaction, the role of the participating members of staff, the activities and individual and collective learning that are generated between participants of a group. To visualise the potential that this concept offers the study, it is necessary to consider the work of Vygotsky (1982) and what is often termed a sociocultural approach that theorises culturally mediated co-construction (Moll 2013). This paper follows an analytical focus on the dialogues and exchanges that occur between the lecturers and the role played by socio-cultural tools or artefacts (such as new forms of talk) in their development.

Vygotsky placed the concept of mediation at the centre of his theoretical thesis. In Vygotsky (1978a) he argued that 'a person undertakes activities through the mediation of others. All of their behaviour is absolutely fused and rooted in social relations' (1978a, 76). In the case of this paper, reference is made to this type of mediation as it focuses on the movement from ideas which develop in social interaction between people to ideas that become personal. This is the essence of Vygotsky's general genetic law of cultural development which celebrates the primacy of social interchange in the development of individual cognitive and affective development.

Faulkner, Littleton, and Woodhead (2013) reframe this concept and refer to 'mediated learning experience' when they discuss the impact mediation can have on the cognitive structure. It is not only knowledge that is mediated but also processing strategies, as Faulkner, Littleton, and Woodhead (2013) indicate 'the existence of groups of strategies and repertoires that allow the organism to use this exposure efficiently has an important influence on cognitive development' (2013, 34). Mediation understood as such may be enacted by an expert in the subject of learning or by a lecturer or a partner who has a more advanced level of knowledge and therefore helps facilitate learning to others (see Watermeyer and Montgomery, this volume). Tharp and Gallimore (1988) also make reference to this type of mediation and call it assisted performance. This form of mediated learning can be strengthened by increasing opportunities for interaction and dialogue. These relationships have the function of mediating the meaning of the world in terms of experiences, perceptions and ideas about practice (Davis et al. 2015). In this way, the subject, through mediated activity in a sociocultural context and participating with others in socioculturally constructed practices, redesigns their world (Hernández 2004).

Tharp (2005) distinguishes five means of supporting others which can be observed in any expert-apprentice relationship: demonstration, contingency management, feedback, task structuring and language media. In the process of teaching and learning, Sannino and Ellis (2013) point out that the psychological effects should not only be considered but the social context of the interaction as well. Thus, there is a mutual shaping of person and situation that is theorised within this framework. For Vygotsky (1978b), an individual's development cannot be understood independently from the social environment in which a person is immersed. The activity context is a scenario where the interaction occurs; this is why it is more than just a physical environment.

In this study, we adopt the notion of context proposed by Gallimore (2002), which 'not only includes aspects of the environment but also the scripts, plans and intentions of the actors' (2002, 41). Therefore, the ways in which interactions occur and the roles taken by people with their representations and meanings is understood as a key part of the context. It is the place where the collaborative interaction occurs and arises from the pressures and resources of the social system of which its actors are part. In this conception, the cultural, environmental, social, cognitive aspects and particularities of a person or group are brought into a dynamic or dialogic relation. In this perspective, the role of the tutor ought to go beyond the mere transmission of knowledge and should be attributed the role of mediator or facilitator in the construction of understanding. Gallimore (2002) claims that without this context, an ideal of development based on mediation cannot be achieved. Vygotsky's (1978b) widely discussed Zone of Proximal Development (ZPD) is a dynamic region which may be thought of as a notional space in which external influences, such as teaching, may, if correctly positioned, be effective in supporting individual progress. The ZPD is in part conditioned by the culture in which the person develops (Moll 2013), as they participate in problem solving along with other members of his cultural group who have greater experience (Davis et al. 2015). A breakthrough in the ZPD is only achieved when that niche provides joint dialectical activity for both the expert and the learner.

In sum, contexts are not to be understood as definitely given, but are formed in a dynamic way, that is mutually shaped by the activity of the participants. In other words, they are constructed from what people do, where and when they do it and, in this sense, the people who interact get to form the context for others (Silvestri and Lacasa 2001). In these spaces, knowledge is constructed jointly as individuals engage in activities that are negotiated rather than imposed.

The notion of the Teacher Support Team (TST)

A TST, like other similar conceptions such as communities of practice, exists in the plurality of social organisations. One can talk about teams in multiple situations, especially when people hold common interests and purposes, share certain belief systems, ways of seeing and doing, and establish various forms of social relationships (groups, associations, organisations) that provide them with support, recognition and identity, while, as a counterpart, they also demand certain commitments and loyalties (Engeström 2014). A TST is a set of people (usually 3 to 5) who meet at a given time and place, engage in a task that requires them to assume roles and interact to achieve a goal. Teams are not an end in themselves; they are a means through which the growth of its members is supported. These groups are not static, but advance in spiral form depending on the task being performed. The interaction that takes place generates development and a specific and group reality that relates the social structure to the individual. For the continuous process of a learning community's development, certain requirements are identified:

- Clear specification of the task to be carried out and mobilisation of the efforts and interests of all to achieve goals and objectives.
- As part of the norms, attendance and punctuality of the programmed sessions.
- Distribution of responsibilities within the group; in other words, the assignment and/or assumption of functions that are rotated from time to time.
- Full participation of all members in constant decision-making and exchanges while in compliance with standards and principles, which can result in the satisfaction of the learning needs of each one of its members and of all as a whole.

Tharp and Gallimore (1988) characterised the nature of these teams as a Joint Productive Activity (JPA). This term was adopted directly from Vygotsky (1978b) suggesting that psychological activity and phenomena depend on and support each other. This perspective is oriented to the understanding of historically defined practices; it is based on a dialectic of knowledge centred on the creative potential of human cognition and seeks to explain the qualitative changes that occur over time in human practices to influence them. The activity is developed through constant negotiation, orchestration and struggle between the different goals and perspectives of the participants. The purpose and motive of a collective activity are like a mosaic in constant evolution, a pattern that never ends completely. In this line, Engeström (2014, 69) notes that:

> Artefact-mediated object-building [...] is a collaborative and dialogical process where different perspectives [...] and voices [...] collide and merge. These different perspectives are rooted in different realities that continue to coexist within the same context of activity.

Gallimore (2002) agrees that the unit of analysis in a setting such as a TST is a joint dialogue and not an individual activity because what matters is the process of social transformation. Tensions are considered forces that motivate change and development. Transitions and reorganisations in the learning team are part of their evolution, which modify not only the mediated activity but also the environment.

Daniels' theoretical construct of TST (1998) is important, especially when we consider that lecturers are very often working on their own when dealing with complex educational and instructional problems. The demands of teaching in general, but also the demands of the continuous, almost non-stop changes that lecturers have to cope with, including attention to diversity, can provoke professional stress and burn-out, leading to lowered tolerance and less personal involvement when lecturers are faced with any situation that call for dealing with new challenges. A TST is explicitly designed to offer a forum where educational professionals can share, discuss and solve immediate and individual problems that arise in lecturers' day-to-day work in the classroom.

Methodological design

It is important to emphasise that one of the purposes of this study is to understand and articulate how university lecturers' knowledge is constructed through dialogue in peer interaction within the setting of a certain activity. The case for the research is a private university in western Mexico which encourages its lecturers within their disciplinary areas to collaborate on the generation of strategies and proposals that will help to improve teaching practices and thus students' learning. The context of the research involves an academic programme, the Bachelor's degree in Engineering, with five disciplinary areas: manufacturing systems, quality systems, process optimisation, project development and supply chain. Thirty adjunct and six full-time Engineering and Science lecturers teach on this academic programme, which enrols 460 students each year.

The units of analysis in this research project are the interactions that take place among the lecturers in the subject of Operations Research (OR), where students learn to use models, statistics and algorithms for studying real complex systems in order to optimise an operation in a certain workplace context. This course is the first of several that aim to help students acquire problem-solving competencies in a business setting. Six university lecturers teach it; their academic profiles indicate they are professionals with up to 15 years of experience with a clear grasp of the subject they are lecturing. In view of the purposes of this research, the decision was made to invite the six STEM lecturers in this area to form a learning team that would come up with learning strategies to help students solve concrete problems that come up in the workplace, i.e. acquiring one of the key competencies that industry looks for in university graduates.

Thus, the decision was made to form a 'Teacher Support Team' (TST). The six lecturers who volunteered from the OR Academy were invited to the first meeting, and told that the researcher's personal interest was to improve teaching practices and thus students' learning results. It was made clear that the study would not focus on evaluation, supervision or control, but rather on the possibility of creating an opportunity for lecturer development, where they would have the chance to examine better ways to promote learning by exchanging experiences and knowledge. It was explained that the meetings would be weekly over three semesters, and would last 90 min each.

The six lecturers agreed to participate in this learning experience, expressing their preference for working autonomously and for distinguishing between professional development meetings and those held to deal with matters related to institutional guidelines or policies or administrative matters. The role of the researcher was clearly spelled out to them from the first contact: a participant observer paying close attention to the interactions among the lecturers.

Epistemological vigilance (Miles and Huberman 1994) refers to an alert state of mind that must exist within the research process in order to avoid confusing common-sense knowledge from the researcher's own experience with scientific knowledge generated through a systematic and rigorous research process. In the case of this study, due to the close participation and relationship between the researcher and the TST, it was considered essential to implement the following strategies: (a) research advisory sessions aimed at maintaining the consistency of the work and coherence between the theoretical and methodological approaches and the findings; (b) the elaboration of written agreements with the research subjects regarding the commitments, scope and limitations of their participation and those of the researcher; and (c) a triangulation to reinforce the analysis by using the points of view of the research subjects.

Data collection called for a series of actions involving the complete team which included the choice of techniques, the design of instruments, the on-site gathering of data, and the recording of information. The observation focused on the following aspects proposed by Gallimore (2002), based on his work on the relationships among participants, performance, and roles grounded in Vygotsky's theoretical framework:

(1) The group members' patterns of individual and collective activity.
(2) The participants' ways of assisting one another.
(3) The different roles the lecturers performed in the activity.
(4) The characteristics of the discussions that arose among the group members.
(5) The linguistic codes and genres used during the meetings.

These referents were used to carry out the observation of the twenty work sessions in which the team of lecturers participated. The basic instruments used to record the data were an audio recorder and a field diary. The twenty audio files that were generated in the meetings with the lecturers were transcribed, and used in the data analysis stage. In the three-hour research process for each of the twenty sessions, the categories and subcategories identified in the analysis of the qualitative data was read by an outside moderator (a STEM lecturer not part of the participant group).

The first step of the analysis process was the complete reading of the transcriptions of the contents of each session which gave insight into the data. These preliminary interpretations were modified over the course of the analysis process and offered a first clue to the conclusions of the research. Data reduction–condensing represents the second stage of this process. In order to accomplish this task, topics were deconstructed as a necessary step toward the construction and organisation of meaningful categories. From the Vygotskian perspective, it can be argued that the segmentation process tried to identify the complete essence and intention of what the lecturers wanted to say. Bakhtin (1986) himself refers to utterance as the unit of analysis and thus of segmentation and it is always expressed from a point of view and in a specific context because it encompasses broader issues such as the speaker's perspective, his conceptual horizon, his intention and his world vision. This premise of sociocultural theory, which coincides with the methodological research conducted by Miles and Huberman (1994), was a decisive aspect when it came to segmenting the texts in order to fully comprehend the meaning the group had in mind. The 'cuts' were made when it was perceived that two or more voices came together over a certain topic or object of the conversation, which generally came up on the initiative of one or more participants.

The change of topic or situation helped to delimit the text segments as well as the questions: 'What are they talking about?' 'What are they referring to?' 'What is the key point?' 'How do they interact from there?' It is important to emphasise that for the purpose of achieving successive approaches to definitions of the phenomenon under examination and to explain it from an emic perspective, the text was segmented in parallel to the assignment of 'codes'.

Once the units to be recorded were identified, segmented and codified, it was possible to proceed to a more detailed analysis. The codes used in the first level of analysis were directly related to the suggestion made by Tharp (2005), but the second analysis went past this proposal and established the following minimum requirements for forming a learning team: agreements and non-agreements, beliefs, assistance in execution, roles, and learning, among others. At the beginning, the concepts proposed by Tharp were quite useful; as Kozulin (2000) points out, there can be no perception process without a theoretical and conceptual referent since this referent determines the way of selecting and interpreting empirical material. In other words, the assignment of codes is not a neutral action, but a way of finding new contexts for analysing and visualising the data on the basis of theoretical preconceptions.

It was possible to develop, refine and qualify the theoretical propositions after these data were conceptualised using the sociocultural theory framework in order to help with the five points mentioned above developed originally by Vygotsky's scheme. First, the transcripts of the twenty meetings were analysed and segmented into the themes in the ideas expressed by the group of lecturers. Once the data reduction task was completed, the next step was to classify, or to group, the new codes on the basis of similarities and differences. For example, when the text segments marked with a certain code were brought together, important differences became apparent. In the case of 'Performance Assistance' it was evident that there were 'types': sometimes the lecturers offered an explanation; on other occasions, they modelled; other times they gave feedback or questioned.

With all the texts grouped on the basis of type of help, it was possible to distinguish more precise units of meaning: the amount of data had become more manageable and offered greater clarity and precision. Having all the interactions where 'Performance Assistance through modelling, questioning, explanation, etc.' could be appreciated made it easier to detect in which cases it occurred, why that particular assistance was offered and not some other kind, and above all, what results were obtained with that help. Separate analysis of the recorded units corresponding to a broader topic allowed for new arrangements and relationships that led to more comprehensive categories. On the basis of these criteria and their ordering along a temporal coordinate, the existence of three phases in the educational process could be discerned:

Interactions–agreements–lack of learning

Engineering tutor B: '[...] At first, we tried to self-define ourselves, that is, know where we were going and what we wanted while keeping our objective of staying academic in mind [...]' (Group interview).

Science tutor F: '[...] How can we balance the objective of the project between everyone? Which conclusions can we reach? Which agreements can we make together for the next session? [...]'

Interactions–lack of agreements–lack of learning

Science tutor C: '[...] How can we evaluate our advancement as lecturers in the quality of the work assigned to the students?

Engineering tutor A: What do you mean?

Science tutor C: From what I understand, we do not have to come to an agreement right away about how we are going to evaluate the student's work with a grade, but about the criteria we are going to keep in mind so that we can say what helps my students learn and develop with more competence in their projects this semester.

Engineering tutor A: But do we have time for that? [...]'

Interactions that generate learning

Engineering tutor E: '[...] For me it was really simple. I am an engineer in electronics; I had studied matrices and was only explained the technique, but I was never taught the *why* and the lecturer explained the fundamentals to me and that was a great learning experience [...]'

Science tutor F: '[...] I am also aware that through these interactions learning must be transferable, in other words, what they are seeing in the classroom they can transfer to company or daily life [...]'

These three stages represent the basic process of the continuous interaction within the TST and how the lecturer moves from the basic level to more advanced levels of interaction.

Results

As seen in the three stages aforementioned in the previous section, we can observe that at the beginning of the TST process, lecturers were focused predominantly on individual work on a basic level; however, as the research continued, they were encouraged to interact with each other, therefore defining and attaining common goals that could bring about continuous improvement in teaching practices as a group according to the five aspects proposed by Gallimore (2002). The sociocultural theoretical perspective highlights the advantages to be gained when people have the chance to interact within a TST in educational spaces where lecturers are free to share ideas, discuss difficult situations and jointly solve problems that arise in the day-to-day reality of their classrooms. This section presents the results of this educational experience in two parts: The first describes the stages of the process in which the team of university STEM lecturers voluntarily participated with an eye to improving and transforming their teaching practice. Special emphasis is given to the lecturers' learning, what helped and what limited the acquisition of new knowledge and its application in day-to-day practice. The second part takes a look at the model that emerges from this educational experience, from the group's particular culture and from the lecturers' previous and new knowledge.

It is important to point out that the structure of this section is directly related to the meanings that have been built on the basis of the data analysis. The first three foci of meaning are developed: 'interactions–agreements–lack of learning'; 'interactions–lack of agreement–lack of learning' and 'interactions that generate learning'. Their arrangement reflects the time continuum corresponding to the way the learning played out, i.e. to the patterns identified at the beginning, middle and end of the process undertaken by this team of professionals. Before these stages are presented, it is important to point out that the learning that came under the headings 'interactions–lack of agreement–lack of learning' and 'interactions that generated learning' actually occurred alternately, i.e. it was not that one came before the other, but rather that depending on the situation, the topic of conversation and the lecturer's interest, one or the other might prevail, and thus they belong to a more comprehensive pattern. The dissatisfaction that a group of lecturers expresses with the quality of student learning influenced their decision to participate in an educational experience.

The interactions, decision-making and agreements correlate to at least four of the basic components of any course: objectives, content, methodology and assessment. This sequence gives a specific shape to the educational process in which the objects of learning – both disciplinary and pedagogical – have a role to play, inasmuch as it is through them that the situations of social exchange arise, as well as the moments in which the attitude and interest in a given content determine whether there is a momentary presence or absence of intersubjectivity on which the opportunity to work and learn together depends. The openness, negotiation and agreement reached on the basis of some resonant meaning must contain a minimum of intersubjectivity among the participants, and this is achieved through dialogue and the acknowledgment of problems that are shared and experienced in a common, reciprocal fashion.

University lecturers' professional development that focuses on a specific interest and an affinity of tasks. in this case on improvement in their teaching practice so as to have a favourable impact on students' performance, was what enabled the reconsideration of viewpoints, which were primarily the product of each lecturer's individual experience. The intention to construct jointly was not easily achieved because each participant had to open up his own individuality in order to understand how the other's reality was represented. In this process, it can be observed that a great deal of the learning that is achieved comes from observation of others and dialogue with others. Modelling has proven to be a significant influence in triggering processes of searching out and development. Intersubjectivity and joint construction have allowed them to rebuild their meanings on the basis of their group experiences, to achieve development as individuals and as a group, to generate specific products and to see themselves as better equipped with both professional and pedagogical tools that they can use in their daily teaching practice.

The experience of this learning team from the area of engineering has helped to identify the following elements that emerge from the data and are considered key for understanding and clarifying the constitution of an educational model. For example, lecturers' academic and professional culture has a significant influence on the definition of objectives, the structure and organisation of the agenda, and the activities and interactions within the educational process. Lecturers come into contact with the cultural legacy that is socially organised and legitimated when their own baggage of knowledge proves to be insufficient for responding to their interest in learning. Pedagogical or educational theory becomes meaningful and offers new ways of understanding when previous knowledge is unable to meet the challenge that lecturers are facing in their everyday practice with students.

Their decision to get involved in an educational project was mediated by real and complex situations that they experienced in their day-to-day activities in the classroom. The group's lived experience and felt needs acted as a symbolic mediator that triggered the lecturers' learning at the individual and group levels. Mediation was a fundamental aspect in the participants' learning zones. In the different types of performance assistance, the lecturers found ways to confer meaning in their teaching and to solve in the 'here and now' problems that they had not been able to solve on their own. The processes of intersubjectivity was recognised in two dimensions: the first in the realm of the already known, i.e. in the group's previous experiences that form part of their academic and professional culture. The second dimension is situated in the problem, in the need and in the challenge felt at a group level in everyday practice.

Figure 1 was constructed using participant information from the observations such as dialogue, mediations, interventions and so on, in order to give a visual representation of the factors involved in the dynamic of forming a TST (Daniels 2017), keeping in mind the three previously mentioned stages regarding the learning process of the entire group. So as to best understand this chart, one must realise that the tutor, whose own culture is built on the basis of past experiences, is the main focus. There are two kinds of situations that are faced: known and thus familiar situations, and experiences that represent a problem. In the latter, the tutor has an interest in solving them satisfactorily. Openness, dialogue and the confrontation of ideas take on meaning in the context of the challenge as the lecturer realises that one's own current tools and cognitive baggage are not up to the task of solving these problems. This is when the willingness to collaborate arises. In this process, transitions can be distinguished that in turn reveal movement, changes, negotiations and displacements. This is not a linear movement, but a spiral, because the new cores of knowledge that are constructed in the 'here and now' take their place among the previous knowledge. The educational model prevailing in this experience belongs to the so-called hybrid models in

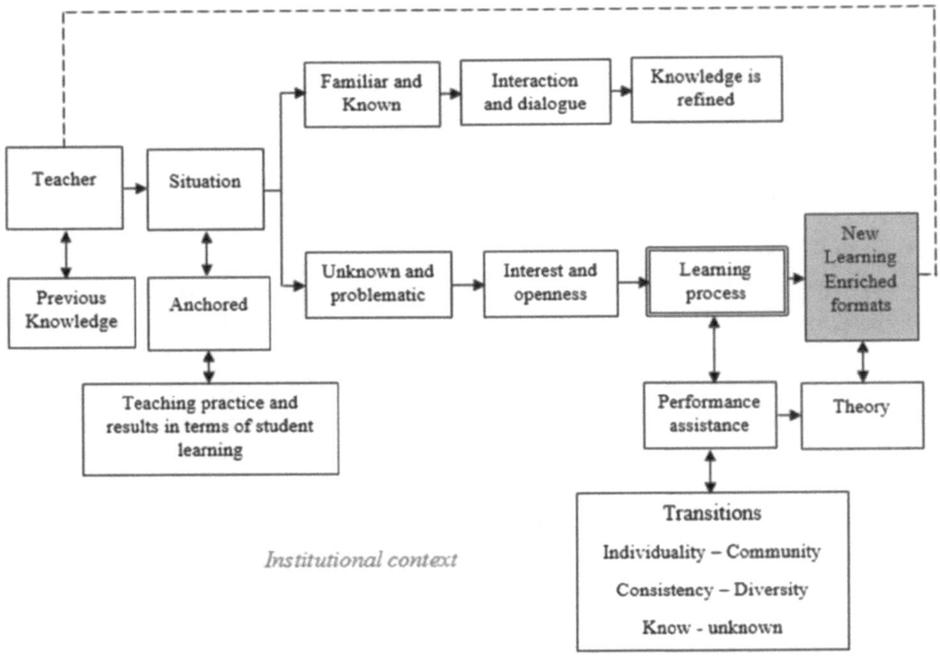

Figure 1. University lecturers' professional development model in a Teacher Support Team.

the sense that the basis for action is predicated on a practical dimension, but without forsaking the theoretical.

Conclusions

Today's world is characterised by generalised accelerated change in the social, political, economic and technological spheres and by a tremendous cultural diversity that has an impact on education in all its elements. It is essential to recognise that the tutor is a core component of the educational process and a bearer of meanings who builds a bridge from the role of the lecturer and practical experience in education to the realm of theory. This section focuses on the clues that might help to achieve an education that is sensitive to the faculty's demand for professional development and to their concrete interests: involving lecturers in educational change can open up new paths to the successful implementation of any educational reform.

A TST in a STEM setting is a new proposal to provide help and support for educators in their practices, especially with the roles and functions of STEM lecturers, who must develop the appropriate collaboration skills, as opposed to the typical profile of one who is accustomed to working alone. The relevance of this study is to experiment with the notion of shifting from individualism to group work. The literature on educational change includes some discussions of university lecturers' professional development programmes that give priority to knowledge derived from theoretical constructions that later are to be translated into applications for different settings and that do not necessarily have a direct impact on the tutor's activity in the classroom.

Their potential for improving lecturers' working conditions means that they offer an important opportunity for intervention, and yet they are not used in universities as much as they might be. They also imply a reversal of the usual learning strategies and relations found in universities in the sense that TST is based on the possibility of sharing knowledge among peers, instead of the typical learning format and hierarchy in which some lecturers act as experts for others. This learning experience with a group of lecturers that set out to build knowledge and improve their students' performance, and decided to collaborate to address situations aimed at enhancing their educational practice, shows that this educational process is a complex undertaking that calls for formal analysis and research in the educational field. This study has made it possible to understand, through dialogue, collaboration and interaction among lecturers, the gap between the intentions at the institutional level where educational models and educational policies are implemented and what lecturers actually do with these guidelines in the classroom.

It is clear that an institution can propose actions, intentions, models and mediations with the idea of supporting lecturers in their efforts to understand educational reforms and restructure their role; however, it is also important to recognise that lecturers' interests do not necessarily coincide with those of the institution because meanings can be very different regarding what needs to be changed in their role, how to achieve this change, and how this might impact learning activities. Appreciating this intricate web of factors that includes perspectives, understandings and actions opens the door to a reassessment of the role lecturers play in this educational reality, in the way the context is understood and interpreted, and in the construction of learning. This experience demonstrates that joint work in modifying and improving teaching practice does not happen by decree; it is driven by a need that lecturers themselves feel in their day-to-day activities. The findings obtained in this study provide ideas for broadening and enriching proposals for university lecturers' professional development, and especially for giving priority to those that take concrete problems from everyday classroom experience and make them the centre of development, the core material that sparks dialogue and gives meaning to collaboration.

This research offers insight from the dialogues and participation that took place between the lecturers when they were provided an opportunity to talk to each other about how professional development can be seen as a personal resource, which is produced out of experiences of their own practice. They can integrate these into more comprehensive processes in which they combine previous knowledge with elements of educational theory, strengthen their role as lecturers from a more open perspective, value mutual support mechanisms, focus on diversity and the unknown, and have a more complete vision of their interventions with students and colleagues. It is important to recognise that the elements of academic culture manifested in lecturers' vision and action with respect to teaching and the role of students in their learning process are vitally relevant in any educational model. If this socio-cultural aspect is neglected, tutor development will be built upon ideals with little likelihood of having any real impact on classroom practice.

Disclosure statement

No potential conflict of interest was reported by the authors.

ORCID

Elvia Castro-Félix http://orcid.org/0000-0002-6428-5477
Harry Daniels http://orcid.org/0000-0003-4523-0670

References

Bakhtin, M. M. 1986. *Speech Genres and Other Late Essays*. Austin: University of Texas Press.

Daniels, H. 1998. *Creación y desarrollo de grupos de apoyo entre profesores* [Creation and development of Teacher Support Teams]. Bilbao: Ediciones mensajero.

Daniels, H. 2017. *An Introduction to Vygotsky*. London: Routledge.

Daniels, H., and P. Garner. 2013. *World Yearbook of Education: Inclusive Education*. Abingdon: Routledge.

Davis, S., B. Ferholt, H. Grainger Clemson, S. M. Jansson, and A. Marjanovic-Shane. 2015. *Dramatic Interactions in Education: Vygotskian and Sociocultural Approaches to Drama, Education and Research*. London: Bloomsbury Publishing.

Engeström, Y. 2014. "Activity Theory and Learning at Work". In *Tätigkeit – Aneignung – Bildung: Positionierungen zwischen Virtualität und Gegenständlichkeit*, edited by U. Deinet and C. Reutlinger, 67–96. Wiesbaden: Springer Fachmedien Wiesbaden. https://doi.org/10.1007/978-3-658-02120-7_3.

Faulkner, D., K. Littleton, and M. Woodhead. 2013. *Learning Relationships in the Classroom*. Abingdon: Routledge.

Gallimore, R. 2002. *Closing the Teaching Gap: Assisting Teachers Adapt to Changing Standards and Assessments*. Los Angeles: Ministry of Education.

Hernández, G. 2004. *Paradigmas en Psicología de la Educación* [Paradigms in Psychology of Education]. México: Paidós.

Kozulin, A. 2000. *Instrumentos psicológicos: la educación desde una perspectiva sociocultural* [Psychological Tools: A Sociocultural Approach to Education]. Barcelona: Paidós.

Miles, M., and M. Huberman. 1994. *Qualitative Data Analysis: An Expanded Sourcebook*. Thousand Oaks, CA: Sage.

Moll, L. 2013. *L.S. Vygotsky and Education*. Abingdon: Routledge.

Sannino, A., and V. Ellis. 2013. *Learning and Collective Creativity: Activity-theoretical and Sociocultural Studies*. Abingdon: Routledge.

Silvestri, A., and P. Lacasa. 2001. "Contextos de aprendizaje y desarrollo. Una mirada desde Latinoamérica." *Cultura y Educación* 13 (4): 339–354.

Tharp, R. 2005. *Activity Setting Observation System*. Santa Cruz: University of California.

Tharp, R., and R. Gallimore. 1988. *Rousing Minds to Life. Teaching and Schooling in Social Context*. New York, NY: Cambridge University Press.

Vygotsky, L. 1978a. *Collected Works of L.S Vygotsky*. New York: Plenum Press.

Vygotsky, L. 1978b. *Mind in Society: The Development of Higher Psychological Processes*. Cambridge, MA: Harvard University Press.

Vygotsky, L. 1982. *The Genesis of Higher Mental Functions. The Concept of Activity in Soviet Psychology*. NY: Plenum Press.

2 Developing a material-dialogic approach to pedagogy to guide science teacher education

Lindsay Hetherington and Rupert Wegerif (England)

ABSTRACT
Dialogic pedagogy is being promoted in science teacher education but the literature on dialogic pedagogy tends to focus on explicit voices, and so runs the risk of overlooking the important role that material objects often play in science education. In this paper we use the findings of a teacher survey and classroom case study to argue that there is a gap in the way that science teachers think about the role of materials and that this could be addressed by changes in the theory base of teacher training, augmenting the current constructivist and dialogic theory with the addition of new materialism in the form of Barad's 'Agential Realism'. Our findings suggests that science teachers do not regularly *explicitly* consider the relationship between the material resources they deploy and the dialogic learning taking place. We argue that science teacher training and professional development should pay more attention to the material-dialogic relationships in the learning that emerges in science classrooms.

Introduction

In this paper we argue that teacher education in dialogic pedagogy in the context of science needs to be expanded and developed to become teacher education for a material-dialogic pedagogy. We make this argument through an analysis of interview and observation data from teachers' experiences of a dialogic inquiry-based science education topic as a stimulus to think through a novel theoretical approach. Our analysis shows a lack of explicit consideration by teachers of the interaction between the material nature of science classroom practice and classroom dialogue and suggests ways in which a different approach to science teacher education could address this.

Research suggests that teacher education is crucial to developing effective dialogic pedagogy (Pehmer, Groschner and Seidel 2015). This is significant as education for more effective dialogue is, we believe, central both to the scientific argumentation approaches and the inquiry-based learning approaches which have been emphasised in recent science teacher education (Hetherington et al. 2016). In addition to dialogic pedagogy, much research that has been influential for teacher education has focused on the use of practical work with tools and materials in school science, often using constructivist and social constructivist ideas (Duit and Treagust 2003; Leach and Scott 2003) although arguably understanding the significance of the role of the materials themselves has been largely neglected (Roehl 2012). However, the relationship between dialogue and the material nature of the science classroom has remained largely unexplored.

Theoretical background

Theories of education informed by constructivism and social constructivism have generated strong research programmes in science education around conceptual change (Taber 2006). Wegerif (2011) questions some of the assumptions of constructivism, arguing that Bakhtin's approach to dialogue offers the basis of a distinct theoretical framework. The capacity to make dialogic switches between different perspectives in dialogue is crucial, Wegerif claims, to the growth of understanding in science and this capacity depends on the quality of relationships. Although this dialogic switching to see and feel things from different points of view is not often emphasised in the discourse of argumentation or the discourse of inquiry-based learning we believe that in fact it is crucial to the success of these approaches. Sometimes the emphaisis in educational dialogue research is on dialogue as a fully external phenomenon visible, audible and tangible in the words spoken and the signs manipulated. Ultimately dialogic switching is not simply an external or visible process, although it might have visible correlates, because it occurs in an internal invisible space of possibilities. Interestingly dialogic switching is not only relevant to relationships between physically embodied individual human voices but can also occur with disembodied cultural voices, the voice of 'Mathematics' for example and with physical artefacts such as computer interfaces (Kazak, Wegerif, and Fujita 2015). This extended understanding of dialogic is pre-figured in Bakhtin's gnomic remark 'I hear voices in everything and dialogic relations between them' (Bakhtin 1986, 169). In science in particular material objects and aspects of materiality need to take on voice to be understood. White (2014) is clear that 'dialogic pedagogy is not a case of hands-off approaches to teaching, but instead an opportunity to 'risk genuine encounters with a multiplicity of others'' (White 2014, 229). In science education those 'others' must include material objects.

From a Bakhtinian perspective, knowledge is a phenomenon that emerges through dialogic interactions; a perspective that sits well within a typical model of teacher education through professional dialogue between teachers and trainee teachers and teacher educators. The importance to continuing professional development (CPD) of establishing a dialogic space between teachers has been brought out in recent research (Hennessy, Mercer, and Warwick 2011; Warwick, Hennessy, and Mercer 2011). The dialogic space affords understanding not only by bringing different discrete voices or points of view into relation with each other but also bringing these into relationship with supervenient voices of the kind that Bakhtin referred to as superaddressee or witness voices. Understanding is often afforded not by seeing one's original perspective in the light of the others perspective but by seeing both perspectives as if from the outside where this capacity to see things as if from the outside does not exist prior to the clash of perspectives but is generated by that clash emerging out of the gap of difference between perspectives (Wegerif 2011).

In science education in particular, the dialogic space needed for effective CPD and teacher education needs to include not only the voices of the teachers but also the 'voices' of materials. The idea that materials might enter into dialogic learning is not obvious and requires a new theory if it is to make sense and also if it is to be of use to teacher education. A key role of teacher education is to challenge the implicit theories of teachers and to provide them with new theory as a guide to action.

There are several attempts to account for the relationship between dialogue and materiality in science education and these have implications for the education of science teachers, (Kress et al. 2001; Cowie, Moreland, and Otrel-Cass 2008; Mercer et al. 2010; Chappell and Craft 2011; Paavola and Hakkarainen 2014). However all these attempts do not explicitly break down the binary distinction between human and non-human in which the material is *used* by humans to create meaning. We use Barad's work to explore an alternative that, like Wegerif's concept of 'dialogic space' (Wegerif 2011), begins with the relationship rather than with the separate parts (Barad 2007). In other words, we reject a substance ontology in favour of a relational ontology. Barad uses her expertise in quantum theory to argue that both matter and meaning are produced through what she calls a material-discursive process. She uses key concepts from quantum physics (in particular Bohr's 'philosophy-physics' drawing out the ontological and epistemological stances that emerge from quantum theory's insights into the relationship between apparatus, observer and observed) to make this argument that the material and discursive are always entangled and intra-acting, with phenomena performed as a result of these ongoing intra-actions (note that we have used the neologism 'intra-action' rather than the more common 'interaction' which Barad uses to highlight the mutually constitutive, entangled nature of matter and meaning). We argue that this material-discursive understanding can be used to extend a dialogic account of the production of meaning in science teacher education and CPD in a way that refocuses on the 'voices' of the material.

This paper explores in particular the difference that this theoretical perspective can make in relation to dialogic pedagogy in the context of science teacher education, building on case study data collected within a large design-based research study, 'Science Education for Diversity' that explored dialogic pedagogy as a means of engaging diverse learners with science (BouJaoude et al. 2013; Wegerif et al. 2013; van Griethuijsen et al. 2015). Drawing out the insights that emerge from interacting this materialist theoretical perspective with the empirical data, we put forward a material-dialogic pedagogy for science teacher education. However, before moving on to develop this argument, it is important to be clear about the terminology we are using in discussing materials and materiality with respect to classroom practice and dialogue.

Ingold (2011) distinguishes between materials and artefacts, where material denotes natural objects and artefacts are man-made objects. This distinction is less useful in Baradian theory where both materials and artefacts as defined by Ingold are intra-actively produced by and productive of phenomena (in our case, learning in a science classroom). We therefore use the term material to refer to both natural and man-made objects used in school science classrooms, including text books, whiteboards, technology such as ipads and, importantly, material used in practical science. These materials are likely to be used in the material-dialogic sense which we begin to develop in this article in particular and different ways, depending on their material nature and the way in which they intra-act (to use Barad's terminology) to create learning phenomena in science. As Otrel-Cass and Cowie (in press) argue, different materials are used in particular ways by teachers and students to support their scientific explanations. The argument we develop here for a material-dialogic pedagogy aims to offer theoretical tools to understand this relationship and guide pedagogy. We use the term materiality to refer to materials as they are experienced through material-discursive intra-action, drawing on Ingold's (2011) notion that materiality is experienced through interaction with an object.

Agential realism

Inspired by the physicist Niels Bohr, Barad (2007) develops a material-discursive onto-epistemology in which matter is an active participant in the performance of phenomena. Barad's theory does not assume a separation between matter and meaning but sees both as united in a single entangled reality. Matter is not given separate agency but becomes an active participant in the process of learning. This approach to agency takes into account that the teacher, pupils and materials all actively make a difference to learning. Rather than understanding teacher agency as 'the capacity for autonomous action' (Calhoun 2002 cited in Biesta and Tedder 2007, 135) or a relational effect between individual and environment (Priestley et al. 2012) where the environment *mediates* the realisation of agency, Barad's work suggests that the agency of the teacher, student and material all *come to exist* in the performance of teaching and learning.

For Barad, the primary ontological units are not things but phenomena, which emerge through material-discursive intra-actions. This is a move away from a representational approach in which language is used to denote or represent, to a performative approach where knowledge and the world are materialised through material-discursive practices. Within agential realism, agency is understood in a relational sense, not as some kind of capacity or power aligned with human intentionality but as 'the ongoing reconfiguring of the world' (Barad 2007, 170); it is 'a matter of intra-action; an enactment, not something someone or something has' (Barad 2007, 177). This implies that matter has a role to play in agency. However, to say that non-human things have agency, as Latour seems to (Latour 2005), would be to misunderstand Barad and to maintain the human/non-human binary which she seeks to overcome. Whilst there are some similarities between Latour's argument in the context of Actor-Network Theory and Agential Realism there are also differences. Latour is very clear that objects are actors within an assemblage but in a distinct way to human actors. For agential realism, by contrast, agency is not a capacity that things might have so things are not directly agentive. Agency is rather understood as something enacted performatively as material-discursive practices make agential cuts that close some possibilities whilst holding open others.

Applying this theory to science classroom practice we can see that that the teacher, students and experiment all participate in learning by making 'cuts', by making some things possible and not others *in the way they relate to and intra-act with each other.* It is this that makes the discursive aspect of Barad's approach important (and in the science classroom, highlights the role of dialogue whilst maintaining a focus on the material).

Barad builds on the insight from research into quantum level reality by showing that the way in which experiments are designed does not simply enable observation of a pre-existing reality but actually brings apparent reality into being. In the famous two slits experiment that is often used in classrooms, one way of observing produces light as particles while another way of observing produces light as waves. For Barad then, apparatuses (including but not restricted to scientific apparatuses) are not simply given to us as things in a world but are ontologically primary mechanisms for creating particular worlds within which they then appear, retrospectively, as phenomena. Apparatuses are open-ended boundary-producing practices which enact agential 'cuts': they are both productive of, and part of, phenomena (where phenomena are both matter and meaning). Wegerif presents dialogic space as a similarly ontologically primary concept, that is, not as an actual space but an opening of a potentiality out of which new configurations of actual spaces, times and meanings emerge within dialogues (Wegerif 2011). Although previously the focus of dialogic theory in education has been on the interaction of human voices, the concept of opening a dialogic space goes beyond this. Dialogic space includes the interaction of 'voices' that are not human. Entangling the ontologically similar concepts of dialogic space and apparatus suggests a way to understand how matter is always already part of the dialogue and how dialogue is also always, in a sense, a material configuration.

Lenz-Taguchi (2011) has applied Barad to education, developing what she calls an intra-active pedagogy, focusing on using pedagogical documentation within material-discursive practices of teacher education. This intra-active pedagogy pays particular attention to the role of pedagogical documentation in enacting agential cuts that both enable and constrain learning. This focus on the role of documentation (texts and artefacts) can be used also to illuminate teacher education where material documents like writing frames for lesson plans and teacher appraisals also enable and constrain the enactment of agency.

Dialogic pedagogy in teacher education

Dialogic pedagogy teaches student teachers how to engage in dialogue for learning together as well as teaching content matter through dialogue (Warwick et al. 2016). What makes any pedagogy dialogic is the aim of teaching and learning for dialogue as an end in itself, in addition to the aim of teaching knowledge. Dialogic pedagogy in teacher education offers a vision of drawing student teachers into participation in continuous professional education as a form of shared inquiry (Warwick, Hennessy, and Mercer 2011). Up to now, although dialogic theory does indeed allow for the engagement with the material (such as the use of text books as an 'other' in dialogic space), presentations of dialogic pedagogy for teaching and for teacher education have not engaged effectively with the material aspect of dialogues (for example the use of key documents to make agential cuts as described above). To investigate the role of the material within a dialogic approach to teaching and learning in science, we have re-analysed data collected as part of the Science Education for Diversity (SED) project (Van Eijck 2011; BouJaoude et al. 2013) alongside new case study data.

Research design

The empirical data discussed in this paper are drawn from an EU-funded international study that used design-based research (DBR) (Barab and Squire 2004; Anderson and Shattuck 2012) to explore theory-driven science education pedagogies for diverse students, following a large-scale exploratory review of international science curricula (Morgan et al. 2012) combined with a survey and field interviews with teachers and students (Van Eijck 2011). A DBR approach was used as it recognises the need to test theory in practical settings using a range of methods to triangulate multiple sources of data within each context to document the enactment of theory-driven pedagogy (The Design-Based Research Collective 2003). Broader findings from the SED project, including questionnaire and interview analysis, are reported elsewhere (see Van Eijck 2011; BouJaoude et al. 2013; van Griethuijsen et al. 2015). Table 1 outlines the SED project research design: in this paper we use data drawn from Work Package 3 (WP3) as a broad context, but focus mainly on data from the DBR element in the UK (WP5). The case study aspect of the project which forms the main data-set for this paper explored the use of a dialogic, place-based pedagogy to engage diverse students with science, drawing on Wegerif's theoretical development of dialogic teaching and learning.

For the purposes of this paper, we draw on the international survey and interview findings to contextualise the importance of practical experience for teacher science education and in addition use data from one of the UK case studies which we have recoded inductively to explore in depth the implementation of the dialogic pedagogy with respect to our focus on the role of the material in dialogic pedagogy and teachers' perspectives on their practice with implications for teacher development. Since DBR is a cyclical process involving long term relationships between researchers and context (The Design-Based Research Collective 2003), we also returned to the setting after 4 years and conducted a further 5 individual, semi-structured interviews with teachers (3 of whom had been interviewed for SED, 2 were current trainee mentors, and 1 trainee) to explore longer term perspectives on the dialogic pedagogy introduced in the project. SED WP5 interviews used a semi-structured format, with a relatively structured interview protocol to ensure consistency between partner countries, then follow-up in a more open-ended style towards the end of the interview; the later interviews used a semi-structured interview schedule that included a greater focus on both dialogue and materiality than the original data-set. The perspective on dialogue as coded in these interviews drew on Wegerif's development of dialogic theory and thus includes a Bakhtinian perspective on dialogue although teachers tended to interpret and use 'dialogic' and 'dialogue' in a sociocultural sense. Interviews were analysed by a process of open line-by-line coding of each individual interview, followed by axial coding and the identification of key themes, which were then checked iteratively by re-coding each individual interview.

Findings

After we briefly outline the results from the large-scale international survey, which highlights the relevance of the role of material resources in science education we focus on the analysis of interviews with teachers bringing out material relevant to teacher education. This was augmented by observations but we do not have space to present that data here. Our analysis reveals that teachers in this study tended to see materiality in two ways depending on context: in the context of information and communications technology they see materiality as a support for learning through dialogue but in the context of the practical work that is a key feature of science education, they see the material as a way of illustrating concepts (there was limited discussion of text books in our data, perhaps reflecting the ethos of the case study school). The potential of dialogue stimulated by/with the materiality of objects for creative learning and growth in understanding is not surfaced in our data; something mirrored by previous research into teachers' purposes for practical work (Abrahams and Saglam 2010). We suggest that this lack of emphasis can be traced back to teacher education and could be overcome, we argue, by a revised teacher education based on a material-dialogic pedagogy.

Table 1. Science education for diversity project methods, sample and analysis.

SED work package (WP)	Methods	Sample	Analysis	Secondary analysis for materi-al-dialogic study
3 Partner Countries: UK, India, Malaysia, the Netherlands, Turkey, the Lebanon	Survey Interviews	9171 students; 331 teachers Focus group interviews with 192 students Individual interviews with 75 teachers	Deductive coding using a coding scheme generated from analysis of pilot interviews with a focus on diversity, to ensure international consistency	No further analysis, conclusions drawn from the primary analysis were used
5	Observations, analysis of classroom materials, interviews and questionnaires	2 schools per country For the case study in this paper: rural English comprehensive school. 2 classes included in the sample, 1 class of 28 students aged 12–13 and 1 class of 25 students aged 14–15. All pupils participated in focus group interviews ($n = 53$), and 2 teachers were interviewed	Inductive coding of qualitative data using constant comparison method of open coding followed by axial coding within each case in each country, then across cases within each country before thematic synthesis between countries	Revisited case study school, re-interviewed 3 teachers and interviewed 2 for the first time Re-analysed original interviews alongside new interviews using open coding and axial coding

Survey findings on use of materials in science education across countries

The international survey of pupils and teachers found that a range of material resources played an important role in science lessons, with differing emphases depending on the distinct national contexts. All countries included mention of practical work, with India, the Lebanon and the UK making most use of standard lab equipment, whereas text books were less commonly used in the UK and the Lebanon compared with other countries (Van Eijck 2011). This variability in the use of materials within science settings is of interest in considering the relationship between the material and dialogic pedagogy and the extent to which science teacher education addresses this relationship.

Interview findings

Here, we report the relevant findings from an inductive thematic analysis of teacher interviews from one of our case study schools, a rural secondary comprehensive school in the SW of England. One of the key findings was that teachers think quite differently about two forms of materiality, the materiality of practical work with objects and tools is thought of in terms of illustrating concepts whereas the materiality of whiteboards and other technology of communication is thought of in terms of opening and supporting dialogue. This gap needs to be addressed by teacher education and is the stimulus for our proposal of a teacher education approach using agentic intra-action from Barad to support teachers thinking about science pedagogy in a new way that includes the voices of materials in the dialogue.

Teachers talking about dialogue and materials

Teachers described their approach to dialogue in terms of the exchange of ideas between pupils, enabling them to try out ideas and experiment with their thinking.

> Teacher F: You've got to show kids that sometimes first of all there isn't necessarily a right or a wrong answer and that there is a place to be creative and there is a place to be brave.

They also identified an exchange of ideas between pupil and teacher within a dialogic pedagogy as fostering an approach where differences of opinion could be highlighted and valued, but also as an opportunity to highlight to pupils the role of evidence in developing an argument in science, thus using a dialogic approach to develop pupils' understanding of the nature of science.

The use of stimulating topics such as natural selection and radiation, were identified by teachers as important in engaging pupils successfully in debate and discussion. Stimulating discussion through raising big questions, or by using an exciting demonstration or practical hands-on activity showed that teachers clearly considered both their questioning and verbal interaction with the students and the use of particular materials within the lesson as important teaching tools. This shows some implicit thinking in relating the material classroom experiences to the exchange of ideas through stimulation of class debate, but, as shown later in the findings and discussion, this relationship is rarely made explicit.

> Teacher B: I like teaching evolution...I like the debate to start with, I like to set it up as a debate and then we'll look at the Darwin's theory and your little jokes about evolution and stuff and you can really get the class engaged with thinking about, you know, 'if we evolved from monkeys, why are there still monkeys?' and you can just start that as

a lesson and have them think about it, and the arguments you get in the class, the quite heated debates are quite nice.

Teacher F: I like teaching radiation because I think it's an exciting subject. It's something that students come across probably for the first time at secondary school. It's reasonably visual, it's got a danger element to it, [and] it stimulates discussion.

This hints at the stimulating nature of the practical work available within these topics, although this link was not specifically made by Teacher F.

The use of film clips and the interactive whiteboards as a basis for prompting discussion, thinking through ideas and questioning was mentioned by multiple teachers in the case study (a finding that was also noted in the broader international interview data-set). This is one of the few points where the teachers made explicit reference to some form of material resource in relation to dialogue.

Teacher C: There might be something on the whiteboard that they would, a question or an image, something that would start them thinking about the topic for example, so they might discuss that in groups and then answer questions.

To summarise, teachers used dialogue in the classroom in tandem with material resources to promote debate and discussion, encourage questioning and a scientific approach, and to enable them to elicit and be responsive to pupils' ideas. Although the materiality of the subject and the classroom is hinted at in relation to dialogue, in our data there was little evidence of clear pedagogical thinking about the role of these materials in the dialogue. It appears that teachers clearly made use of them to stimulate and extend discussion but little was said about *how* this took place. In order to develop teachers' dialogic pedagogy as part of professional development, we would therefore suggest that professional dialogue in a teacher education or professional development programme begins with a focus on surfacing how teachers use dialogue in the classroom in and of itself and in relation to the classroom materials and environment. The concept of agency as developed by Barad suggests that the intra-action of teachers, students *and* materials are all crucial to the performance of learning, in this case in the science classroom. The need to be responsive, as identified by the teachers in this data, is crucial to this agentic intra-action but the focus in the data remains on teachers responding to students, with little consideration of the material. This reflects findings elsewhere by Roehl (2012), suggesting that due to the ubiquitous nature of materials in Science, *how* they are used in science teaching and learning tends to be overlooked. This gap could be addressed through teacher education through the use of teachers' pedagogical and classroom documentation and dialogue in a fashion that models the intra-actions in the science classroom.

Teachers talking about their use of material resources in science education

Teachers made use of a range of material resources in their lessons and made regular reference to the use of science-specific resources on the one hand and technology on the other. The use of materials and modelling to construct ideas was mentioned. This is interesting in the context of this paper as it is one potential aspect of science teachers' common practice that could be drawn upon to develop their understanding of material-dialogic relationships.

However, although science-specific material resources were seen to be important, it was, as noted above, rare for teachers to make explicit *how* they used these materials in relation to dialogic pedagogy or to stimulate pupils' thinking. One exception to this was Teacher A, who participated in the Science Education for Diversity intervention and was concerned to ensure that practical lessons were used to explore ideas.

> Teacher A: Even if I was doing a practical I would make sure that was interspersed with some sort of coming together, exchange of ideas.

Teachers in this case study school all valued and made use of technology. They felt that technology could be used to stimulate thinking and prompt discussion, as illustrated by the quote from Teacher C above, and Teacher E's statement that he 'uses technology, and media input to get them thinking about things'. Teachers regularly noted that a lack of technology could be an obstacle to learning, and were keen to use cutting edge technologies in their lessons.

Interestingly, there was more regular consideration of how technologies such as iPads, interactive whiteboards and film clips could be used to stimulate thinking and in the context of class or group discussion than when teachers discussed the use of scientific practical equipment. Teacher education and professional development has been more explicit in linking these aspects, with practical work being more likely to be discussed with reference to conceptual illustration and development and scientific inquiry and investigation as a general pedagogical framework than explicitly dialogic.

> Teacher E: Very briefly and as simply as possible, [good teaching is] putting the learning in the pupils hands, giving them materials and resources for them to construct their own ideas.

> Teacher B: I've got my year eights marching on the playground to do refraction of light, just teach them to march very, very quickly, 'when you cross this line you have to slow down, hit it at an angle the whole lot moves' and you actually pupil model the refraction of light really, really well.

Modelling has long been an important aspect of science and science education, with teaching incorporating the use of scientific models and the notion of modelling as an important aspect of the nature of science, alongside the use of both physical and mental models as teaching aids (Gilbert and Boulter 1998). This important element of science pedagogy is often a core concept included in science teacher education and it is therefore no surprise that teachers mentioned the use of models to aid pupils in understanding concepts and constructing ideas. As the two quotes above show, teachers are keen to use materials, including pupils' own bodies, to help them model scientific ideas in order to understand them. Again, of interest in relation to this study, teachers regularly mentioned the use of various kinds of models for this purpose, but did not explicitly link their use to classroom dialogue; they did not tend to explore or explain how they could use dialogic pedagogy to work with pupils to construct models or use models to construct their own ideas about science. This, it seems to us, is one locus of potential in supporting teachers to further develop their practice in material-dialogic pedagogy as there is already an implicit relationship between material, model, classroom talk and scientific concept.

Teachers talking about teacher education in relation to dialogue and to materiality

The teachers in our case study discussed teacher education in relation to the importance of good relationships with pupils, and being able to use practical work in particular effectively, firstly in a logistical sense and then in the sense of stimulating thinking. Thus, they focused primarily on practical work with respect to teacher learning about interaction with science classroom materials.

The teachers in the sample regularly noted the importance of good relationships with the pupils in order for dialogic teaching and indeed, all science teaching, to be effective, and also highlighted this aspect as something that is crucial in teacher education. For example, Teacher E felt strongly that the 'single most important thing in teacher training…is [for trainees to] be aware that good teaching is based on a relationship with first of all the group, then groups, then individuals'.

The trainee teachers' mentors (these were experienced teachers within the school who support trainee teachers in the classroom and shape the teaching experiences they are offered) tended to note that new teachers needed supporting with the use of practical work, initially in a logistical sense before they could consider how to use the practical most effectively to stimulate learning.

Teacher C: So he [the trainee] might say 'so what we need to focus on isn't necessarily the pedagogy of using practicals to help learning but it's more about the logistics of practicals in the room'.

Teacher G: [The trainee] misses out on capitalising on practicals.

Interestingly, this point demonstrates the trainee teachers developing pedagogical skills and mirrors Abrahams and Millar's framework for effective use of practical work through effectiveness at the level of observables and the level of ideas (Abrahams and Millar 2008). In their study, Abrahams and Millar noted that practical work was often effective in terms of making observations and completing the task, but less often effective in supporting students' developing understanding. Further, the teachers' discussion, both in our findings and in Abrahams and Millar's work, does not place a clear and explicit focus on exploration of the role of dialogue in relation to the practical work (i.e. the use of materials) and how they work together. This suggests that teacher education for a material-dialogic approach could aid teachers in making more effective use of practical work despite the constraints they face in terms of curriculum time and resourcing (SCORE 2008).

In our data, mentors and trainee teachers made little reference to conversations between them about the way dialogic pedagogy could be used in relation to practical work or other materials such as the use of interactive whiteboards. Given the time constraints they face this is probably not surprising, but it is crucial because the mentors are the key point of contact for trainee teachers undertaking their training in school, observing them teach and discussing their teaching with them with a view to improving and extending their practice. One teacher in the case study school noted that they let the pupils discuss during practicals but they did not focus on scaffolding those discussions:

Teacher G: It's sometimes very interesting to listen to their conversations when they're doing practicals…I suppose it gives them something solid to talk about.

This teacher went on to explain that she lacks confidence in running discussions, noting that 'I'm not particularly good at saying, 'go away and talk about this', perhaps I don't set it out right; the conversation drifts'. This teacher was herself involved in teacher education as the mentor to a trainee teacher in the school. It is evident from this and other comments by the teachers that there is a need for more teacher education and continuing professional development about how to make effective use of dialogue and dialogic pedagogy in relation to classroom practice in science, both generally and with respect to materials in a science classroom context.

Discussion

The interview findings show that material and dialogue both play a clear role in science classroom practice. However, although both elements are valued and (using a material-discursive framing) are intra-acting to perform or enact learning, our data showed little explicit consideration by teachers of the interaction between the material nature of classroom practice and the dialogue that takes place. The dialogues are material through words, signs, cards, templates and practical demonstrations, which all both constrained and enabled learning, making a difference (and making new material configurations) that open up new material-dialogic spaces, but this intra-action of material and dialogue was not surfaced by teachers in this study. Our findings also show how, mirroring this classroom practice, the trainee-teacher and teacher mentor relationship within school (a key part of teacher education in the UK) has both material and dialogic elements but that there was implicit but not explicit consideration by teacher mentors of material-dialogic interaction either with classroom materials or teacher education pedagogical materials in the performance of teacher learning. The importance of relational interactions between teachers and pedagogic materials in teacher learning has previously been highlighted by McNicholl, Childs, and Burn (2013), who argued that teachers' pedagogical content knowledge could no longer be viewed as individualistic but rather distributed via artefacts held in school science departments, with practical implications for teacher education in the importance of contribution to shared pedagogical materials and the need for a shared physical space for teachers to collaborate. However, they conceptualise the material and dialogic spaces separately.

The separation commonly found in the literature between dialogic space and the voices participating in learning, and the material world, was also highlighted in the teacher interviews; the material was valued as offering stimulating practicals, enabling modelling and illustration of concepts, and in the use of technologies to stimulate thinking and questioning, but there was no clear articulation by these teachers about the relationship between the material, the dialogue and the learning. For example, the role of evidence in scientific argumentation and dialogue is one aspect where, although the material nature of that evidence was not made explicit in the teacher interviews, materiality may have a key role to play. Similarly, Teacher B's comment about the stimulating nature of the visual demonstration associated with teaching about radioactivity demonstrates that although teachers were clear that both material resources and dialogue helped to stimulate pupils' thinking, links between material and dialogue appear to be implicit rather than explicit in the teachers' thinking. From the comments made we can deduce that this separation of dialogue from material stems from teacher education and could be addressed through changes in teacher education.

Drawing together a Bakhtinian dialogic pedagogy (Wegerif 2008) with Barad's material-discursive onto-epistemology (Barad 2007) may offer a theory that could underpin a teacher education approach that would take a more dialogic approach to relationships with materials. Looking at the data using this lens, we can see that teachers, students *and* materials enact agential cuts through their intra-actions that enable particular learning phenomena to emerge. An agential cut is a particular observation that divides reality making some aspects visible and others invisible. In the classroom we see how discourse and matter conspire to produce key decisions that shape learning. In collaboration between science teachers to develop their pedagogical content knowledge such as studied by McNicholl, Childs, and

Burn (2013), the concept of the agential cut could also be useful in exploring *how* the space, pedagogical artefacts and teachers intra-act to enable particular learning.

Significantly, the distinction between a pedagogical approach influenced by Barad's agential realism and other dialogic pedagogical approaches, is in the understanding that the material has agency in enacting agential cuts. The argument goes beyond the suggestion that science teachers should pay attention to the material in their lessons. This is evidently already the case. What we are arguing for is a pedagogy in which science teachers are explicitly aware that their intra-action with both the students *and* the material generates distinct phenomena that matter in the world; that the becomings of material-student-teacher (i.e. learning) are enabled and constrained by the choices that are made through those ongoing intra-actions.

In teaching about electrical circuits, teachers could ask themselves what learning is enabled and constrained if the practical materials are offered to the students as a locktronics kit with recipe style instructions to follow, or if they are given the same kit and they are told to play with it and present their findings to the group, or if they are given some everyday objects, wires, crocodile clips and a few cells and asked to make the objects work. This could be supported in itself through the relationship between material and dialogue within a teacher development programme. For example, in the case study school here, trainee teachers are required to use particular forms of documentation: the agenda and the framework for dialogue about teaching (Hetherington, Postlethwaite, and Skinner 2015). These pieces of documentation ask the trainee teacher to make explicit their intentions for a lesson in relation to an aspect of their own development (not the students' learning which is the focus of the lesson plan). They then teach whilst their mentor makes non-evaluative notes against their intentions as outlined in the agenda. The agenda and annotations then become the focus of reflective dialogue between trainee teacher and mentor in a discussion that may be supported with reference to the framework for dialogue about teaching. In the example suggested above, how the trainee teacher frames the electrical circuit lesson could be written into an agenda about the use of practical resources and thus used to explore the material-dialogic intra-action in the classroom. In the case study, this process was not being used to focus on the material dialogic relations in part because mentors were not confident to do so and in part because it was not a focus suggested by the university. Teacher education and development could therefore be extended by the university making the material-dialogic interaction an explicit potential agenda focus and supporting mentors in this through their mentor development programme, and by making stronger use of the agenda tool within trainee teacher placements. Using the pedgagogical documentation of teacher education in this way would also model for trainee teachers and their mentors how material resources can be seen as entering into the material-dialogic space through reflective discussion of how the framing, use and discussion of an agenda makes agential cuts in the learning of the trainee teacher and mentor (and, indeed, the students in the class). Although the range of learning opportunities offered to the students could be standard, the difference here is in the way the teacher *approaches* the lesson and how they plan to scaffold the lesson through material-dialogic interaction with an interest in responsively acting on the learning that emerges as a result.

This awareness of how matter enters into and shapes learning extends the dialogic space. Recognising how a particular tool, a piece of graph paper for example, both constrains and enables learning enables that tool to be interrogated and to enter into the larger dialogue not as passive background but as an active or constitutive voice within the dialogue. Science teachers taking up a material-dialogic pedagogical approach will therefore need to consider how their choices of materials and communication tools intra-act in dialogue with themselves and their pupils in generating emergent learning, and, given that this is not currently an explicit element in science teacher education, we suggest that science teacher educators need to engage with such an approach.

It is apparent from our findings that although teachers value pupil discussion in science and link this to teaching the nature of science, there is not always a strong sense that those experienced teachers who are involved in teaching student teachers are confident in using dialogic pedagogy even in a school involved in practice-based research in this area.

Further, the focus on developing logistical competence *before* focusing on pedagogical rationale suggests that in initial teacher education, dialogic pedagogy could remain a theoretical concern until trainee teachers are competent in managing the materiality of classroom practice. However, teacher education and professional development that will enable teachers to feel confident in using material-dialogic pedagogy will be challenging to implement, if lessons learned from the 'Getting Practical' project, which aimed to support teachers in using 'minds on' as well as 'hands on' approaches to practical work, are considered (Abrahams, Reiss, and Sharpe 2014). A multi-pronged approach to shifting pedagogy that brings material-dialogic theory to practice through university-based teacher education programmes and tools, school-based experiences and collaborative dialogue between teachers, and teacher-pupil-material intra-actions through shifting classroom pedagogy are all necessary elements in bringing these ideas to fruition in practice. This requires further exploration of effective teacher education pedagogy with respect to these ideas as well as time and support for teachers.

We would argue, however, that shifting focus towards a material-dialogic approach would support both teachers and teacher educators in facilitating new teachers to manage both elements of the class simultaneously and effectively. Baradian theory draws the notion of agency into this pedagogical approach in useful ways as it explicitly acknowledges the role of teacher, trainee teacher, pupil and the material in an agentic performance of learning. For example, one challenge for student science teachers is how to differentiate a practical activity. A common default for differentiation is setting book work for those who finish early. A dialogic material-discursive alternative is to allow the students to explore working with the equipment to see what happens when they configure it differently. What applies to students in classrooms applies equally strongly to student teachers within teacher education. Rather than teaching the use of material to illustrate conceptual knowledge student teachers should be taught how to play with materials in order to explore knowledge. This would enable them to experience the potential of working with matter in a way that allows its voice to speak and enter into a dialogue; a way that allows/acknowledges its participation in an emergent shared agency. For dialogic pedagogy the aim is, amongst other things, to expand understanding and the potential for creative action by expanding the dialogic space. In current classrooms too much that pertains to the material conditions and media of science education is taken for granted and so becomes the passive context and frame for dialogue. Incorporating a greater awareness of material conditions of every kind that limit as well as enable dialogues is a way of expanding dialogic space. Simply playing with equipment and exploring alternative configurations of tools is a way of bring the inert background into active play within learning dialogues. We suggest that further research to understand the different ways teachers work with students and different material elements of the science classroom, including but not limited to practical equipment, would help refine and support the implementation of the concept of material-dialogic pedagogy developed here. This pedagogical framework has the potential to transform science teacher education by combining the mastery and use of tools for teaching with the mastery and use of dialogic pedagogy because these are now seen to be just two aspects of one sole process, the dialogic discursive material process of learning.

Disclosure statement

No potential conflict of interest was reported by the authors.

Funding

This work was supported by EU FP7 programme [the Science Education for Diversity (SED) Project FP7-SCIENCE-IN-SOCIETY-2009-1 244717].

ORCID

Lindsay Hetherington http://orcid.org/0000-0002-9811-7666
Rupert Wegerif http://orcid.org/0000-0003-2278-2245

References

Abrahams, I., and R. Millar. 2008. "Does Practical Work Really Work? A Study of the Effectiveness of Practical Work as a Teaching and Learning Method in School Science." *International Journal of Science Education* 30 (14): 1945–1969.

Abrahams, I., M. Reiss, and R. Sharpe. 2014. "The Impact of the 'Getting Practical: Improving Practical Work in Science' Continuing Professional Development Programme on Teachers' Ideas and Practice in Science Practical Work." *Research in Science and Technological Education* 32 (3): 263–280.

Abrahams, I., and M. Saglam. 2010. "A Study of Teachers' Views on Practical Work in Secondary Schools in England and Wales." *International Journal of Science Education* 32 (6): 753–768.

Anderson, T., and J. Shattuck. 2012. "Design-Based Research: A Decade of Progress in Education Research?" *Educational Researcher* 41 (1): 16–25.

Bakhtin, M. 1986. *Speech Genres and Other Late Essays*, translated by V. W. McGee. Austin, TX: University of Texas Press.

Barab, S., and K. Squire. 2004. "Design-Based Research: Putting a Stake in the Ground." *Journal of the Learning Sciences* 13 (1): 1–14.

Barad, K. 2007. *Meeting the Universe Half-Way: Quantum Physics and the Entanglement of Matter and Meaning*. Durham: Duke University Press.

Biesta, G., and M. Tedder. 2007. "Agency and Learning in the Lifecourse: Towards an Ecological Perspective." *Studies in the Education of Adults* 39 (2): 132–149.

BouJaoude, S., R. Khishfe, S. Alameh, and R. Saad. 2013. "Science Education for Diversity: WP5 Deliverable 5.1." Edited by R. Wegerif. Accessed July 31, 2015. http://www.science-education-4-diversity.eu

Calhoun, C. 2002. *Dictionary of Social Sciences*. Oxford: Oxford University Press.

Chappell, K., and A. Craft. 2011. "Creative Learning Conversations: Producing Living Dialogic Spaces." *Educational Research* 53 (3): 363–385.

Cowie, B., J. Moreland, A. Jones, and K. Otrel-Cass. 2008. *The Classroom InSiTE Project: Uunderstanding Classroom Interactions to Enhance Teaching and Learning in Science and Technology in Years 1-8 (Final Report)*. Hamilton: University of Waikato.

Duit, R., and D. F. Treagust. 2003. "Conceptual Change: A Powerful Framework for Improving Science Teaching and Learning." *International Journal of Science Education* 25 (6): 671–688.

Gilbert, J. K., and C. J. Boulter. 1998. "Learning Science through Models and Modelling." In *International Handbook of Science Education*, edited by B. J. Fraser and K. G. Tobin, pp. 53–56. Dordrecht: Springer.

van Griethuijsen, R., M. van Eijck, H. Haste, P. den Brok, N. Skinner, N. Mansour, A. Savran Gencer, and S. BouJaoude. 2015. "Global Patterns in Students' Views of Science and Interest in Science." *Research in Science Education* 45 (4): 581–603.

Hennessy, S., N. Mercer, and P. Warwick. 2011. "A Dialogic Inquiry Approach to Working with Teachers in Developing Classroom Dialogue." *Teachers College Record* 113 (9): 1906–1959.

Hetherington, L., K. Postlethwaite, and N. Skinner. 2015. *The Exeter Model of Initial Teacher Education*. Exeter: University of Exeter.

Hetherington, L., H. Ruck Keene, K. Chappell, and C. Slade. 2016. "CREATIONS: Developing an Engaging Science Classroom. D2.4 the Professional Development of Educators: Consideration and Strategies." Accessed December 8, 2017. http://www.creations-project.eu

Ingold, T. 2011. *Being Alive: Essays on Movement, Knowledge and Description*. London: Routledge.

Kazak, S., R. Wegerif, and T. Fujita. 2015. "The Importance of Dialogic Processes to Conceptual Development in Mathematics." *Educational Studies in Mathematics* 90 (2): 105–120.

Kress, G., C. Jewitt, J. Ogborn, and T. Charalampos. 2001. *Multimodal Teaching and Learning: The Rhetorics of the Science Classroom, Advances in Applied Linguistics*. London: Continuum.

Latour, B. 2005. *Reassembling the Social: An Introduction to Actor-Network-Theory*. Oxford: Oxford University Press.

Leach, J., and P. Scott. 2003. "Individual and Sociocultural Views of Learning in Science Education." *Science and Education* 12: 91–113.

Lenz-Taguchi, H. 2011. *Going beyond the Theory/Practice Divide in Early Childhood Education: Introducing an Intra-Active Pedagogy*. Abingdon: Routledge.

McNicholl, J., A. Childs, and K. Burn. 2013. "School Subject Departments as Sites for Science Teachers Learning Pedagogical Content Knowledge." *Teacher Development* 17 (2): 155–175.

Mercer, N., P. Warwick, R. Kershner, and J. Staarman. 2010. "Can the Interactive Whiteboard Help to Provide 'dialogic Space' for Children's Collaborative Activity?" *Language & Education: An International Journal* 24 (5): 367–384.

Morgan, A., B. Choksi, S. BouJade, N. Swee Chin, R. van Griethuijsen, and H. Bag. 2012. *Science Education for Diversity WP2 Exploration: Documentary Analysis Synthesis Report*.

Otrel-Cass, K, and B. Cowie. (in press). "The Materiality of Materials and Artefacts used in Science Classrooms." In *Material Practice and Materiality: Too Long Ignored in Science Education*, edited by C. Milne and K. Scantlebury. New York: Springer.

Paavola, S., and K. Hakkarainen. 2014. "Trialogical Approach for Knowledge Creation." In *Knowledge Creation in Education*, edited by S. G. Tan and J. Yeo, 53–73.Singapore: Springer.

Pehmer, A., A. Groschner, and T. Seidel. 2015. "How Teacher Professional Development regarding Classroom Dialogue Affects Students' Higher-Order Learning." *Teaching and Teacher Education* 47: 108–119.

Priestley, M., R. Edwards, A. Priestley, and K. Miller. 2012. "Teacher Agency in Curriculum Making: Agents of Change and Spaces for Manoeuvre." *Curriculum Inquiry* 42 (2): 191–214.

Roehl, T. 2012. "From Witnessing to Recording – Material Objects and the Epistemic Configuration of Science Classes." *Pedagogy, Culture & Society* 20 (1): 49–70.

SCORE. 2008. *Practical Work in Science: A Report and Proposal for a Strategic Framework*. London: Science Community Representing Education.

Taber, K. 2006. "Beyond Constructivism: The Progressive Research Programme into Learning Science." *Studies in Science Education* 42: 125–184.

The Design-Based Research Collective. 2003. "Design-Based Research: An Emerging Paradigm for Educational Inquiry." *Educational Researcher* 32 (1): 5–8.

Van Eijck, M. 2011. *Science Education for Diversity Work Package 3 Synthesis Report*.

Warwick, P., S. Hennessy, and N. Mercer. 2011. "Promoting Teacher and School Development through Co-Enquiry: Developing Interactive Whiteboard Use in a 'Dialogic Classroom'." *Teachers and Teaching: Theory and Practice* 17 (3): 303–324.

Warwick, P., M. Vrikki, J. D. Vermunt, N. Mercer, and N. van Halem. 2016. "Connecting Observations of Student and Teacher Learning: An Examination of Dialogic Processes in Lesson Study Discussions in Mathematics." *ZDM Mathematics Education* 48 (4): 555–569.

Wegerif, R. 2008. "Dialogic or Dialectic? The Significance of Ontological Assumptions in Research on Educational Dialogue." *British Educational Research Journal* 34 (3): 347–361.

Wegerif, R. 2011. "Towards a Dialogic Theory of How Children Learn to Think." *Thinking Skills & Creativity* 6 (3): 179–190.

Wegerif, R., K. Postlethwaite, N. Skinner, N. Mansour, A. Morgan, and L. Hetherington. 2013. "Dialogic Science Education for Diversity." In *Science Education for Diversity*, edited by R. Wegerif and N. Mansour. New Jersey, NJ: Springer Science.

White, E. J. 2014. "Bakhtinian Dialogic and Vygotskian Dialectic: Compatabilities and Contradictions in the Classroom?" *Educational Philosophy & Theory* 46 (3): 220–236.

3 Creating a dialogic environment for transformative science teaching practices

Towards an inclusive education for science

Cristina G. Reynaga-Peña, Marisol Sandoval-Ríos, José Torres-Frías, Carolina López-Suero, Adrián Lozano Garza, Maribel Dessens Félix, Marcelino González Maitland and Jorge G. Ibanez (Mexico)

ABSTRACT

This paper focuses on the design and application of a teacher training strategy to promote the inclusive education of students with disabilities in the science classroom, through the creation of adult learning environments grounded on the principles of dialogic learning. Participants of the workshop proposal consisted of a group of twelve teachers who were working at various educational levels. Teacher teams proposed and implemented in their classroom, innovative, inclusive science-learning activities about a topic of their choice. Data were collected from interviews with teachers five months after the courses, teachers' portfolios about their practice during implementation of such sequences, and researchers' observations. The data suggest that it is possible to stimulate a gradual transformation of teaching practices through a teacher training proposal that promotes self-awareness and critical reflection, situated in the creation of meaning and a willingness to change in the spirit of solidarity and social action. We found elements to recommend the incorporation of these innovations at the curricular and practical level for teacher training schools and for in-service teacher development programmes in Mexico and elsewhere.

Introduction

Science education has become a priority for educational systems worldwide in the last decade, while the importance of evolving new strategies for construction and application of contextualised scientific knowledge beginning early in education is also recognised. In Mexico, the status of science education is not very encouraging. In a thorough review of Mexican science education programmes and practices at the primary and secondary school levels, Flores-Camacho (2012) concluded that sciences held a secondary place in the public education system, as the programmes emphasised language and mathematics.

Flores-Camacho also claimed that, as a result, the resources for the primary school level (materials, teacher training and funds) were focused on those areas. According to Flores-Camacho, this situation was partially due to the generalised perception that elementary-level science was a non-essential, complementary area of no use for everyday life, and refer to the idea of science as exclusive for people with superior capabilities. The review also indicates that at the secondary school level, the situation was not much different, where 'the vision of science and science teaching attached to dictation and memory give a distorted image of science, as effectively, marginal and useless' (2012, 6). This analysis took place just before the implementation of an educational reform where efforts to improve education also aimed for better preparation for in-service teachers (Ruiz 2012). Since that reform, a recent educational model was proposed (SEP 2017) where inclusive education and equity are now a priority. Through the changes and reforms, the role of teachers has become more significant. Since the education reform first proposed by Shulman (1987), and currently adopted in many countries, the role of teachers was emphasised worldwide (Villegas-Reimers 2003). Even though inclusion is a multifactorial issue, teachers play a fundamental role in implementing inclusive environments for all children in the classroom.

For these reasons, this research team directed efforts towards a teacher-training programme focused on an inclusive education of science upon the premise that this approach enhances science learning for all students in order to include them into a knowledge society, so they can achieve full personal and social development. Although we are aware that, internationally, science education has been reconceptualised as STEM education, in Mexico the educational policies still feature science as a separate subject.

Teaching science to students with disabilities

A major challenge for science teaching is the inclusive education of students with disabilities and in particular with visual impairments, given that science is saturated with representations of scientific concepts and most of these representations are visual (Jones et al. 2006). Nevertheless, science has also been identified by teachers as the most amenable subject for mainstreaming students with different disabilities (Atwood and Oldham 1985; McGinnis and Stefanich 2006), because it allows students to develop compensatory skills for observing, manipulating, and classifying phenomena, in addition to promoting collaborative learning (Mastropieri and Scruggs 1992). We also used the case of visually impaired students to illustrate inclusive practices in science teaching because nowadays most science teaching in Mexico privileges the use of visual resources (images, videos, and observations with the sight) and we are committed to promoting more engaging hands-on science learning experiences that make use of all senses.

In many countries, teacher preparation programmes are substantially different for primary school, science or special education teaching; therefore, teachers in mainstream classrooms usually do not have training pertaining to teaching students with disabilities, and special education teachers have insufficient or no exposure to science teaching pedagogy (Cawley 1994; Norman, Caseau, and Stefanich 1998). Even nowadays, the Mexican teacher training curriculum does not specifically cover the area of teaching science for inclusion and it is just recently that one general inclusive education course has been inserted into this programme (SEP 2012). It is thus crucial to generate proposals that prepare teachers in how to create inclusive environments, while building science teaching skills and knowledge as the basis for transformation of their practice (NAP 2015). However, in order for this to occur, it is necessary to support teachers in incorporating the value of diversity and to stress the need for creating inclusive practices by reorienting teacher preparation (Booth and Ainscow 2002; European Agency 2010; UNESCO 2009). This can be achieved through working with the principles of dialogism, transformative learning, and the construction of dialogic learning communities.

Dialogic learning and its role in inclusive education for science

According to García (2012, 51) 'effective teaching and learning processes respond to the communicative conception of learning in which dialogue and interaction are key elements'. In recent decades, science teaching presented a shift towards communicative conceptions that have contributed to improving teaching and learning processes with an increased emphasis on social interaction. From this perspective, dialogic learning focuses on the inter-subjective and communicative learning dimension, and evidence a scientific framework to explain how people can learn more and do so better (Aubert et al. 2008). Literature shows that dialogic learning is a key factor of teaching and learning science, with special attention to the creation of knowledge mediated by a reflective and critical dialogue (Molina and Ríos 2010). The concept of dialogic learning is grounded in a communicative conception in which the subject learns from the interaction with other people to promote egalitarian and scientific educational practices (Aubert et al. 2008; Flecha 1997; Flecha, Gómez, and Puigvert 2001). In other words, this is a theoretical perspective that focuses on egalitarian dialogue.

The seven principles that set the basis of dialogic learning aim at an educational and social transformation in which promoting inclusive education of students with disabilities can be achieved, namely: (a) egalitarian dialogue, where all opinions have the same weight in a collective construction act; (b) cultural intelligence, recognising that all participants provide significant knowledge independently of their cultural origin; (c) transformation, as dialogic learning transforms not only existing relationships, but also between people and their environment, as well as their own knowledge; (d) instrumental dimension, involving communication and intersubjective dialogue, seeking to give maximum opportunities to all students, especially those at risk of suffering exclusion; (e) creation of meaning, where learning is developed to create meaning for those involved in the teaching and learning, so knowledge has significance and practical value; (f) solidarity, so that the idea of individual success is substituted for common well-being through relationships based on transcending of individual participants; g) equality of differences, so diversity and differences between individuals are seen as complementary and rich aspects rather than antagonist aspects (Aubert et al. 2008; Flecha 1997; García 2012). The principles that set the foundation of dialogic learning largely respond to the specific needs of teaching students with disabilities, since this framework integrates crucial aspects of inclusive education. Indeed, dialogic learning supports theories, practices, and empirical evidence of educational strategies that contribute to social transformation to overcome the cycle of social inequalities through education (Prieto and Duque 2009).

Learning to participate in a dialogue involves the capability of engaging in reflective judgment (Mezirow 2003), by which assumptions and expectations supporting their own beliefs, values, and feelings are assessed by the individual, so he/she can offer a perspective about his/her own perspective, this being a condition for transformative learning. The transformation theory of adult learning was conceptualised by Mezirow (1997). In his own words, Mezirow described this concept as:

> Learning that transforms problematic frames of reference – sets of fixed assumptions and expectations (habits of mind, meaning perspectives, mindsets) – to make them more inclusive, discriminating, open, reflective, and emotionally able to change. (2003, 58)

When knowledge is constructed through the synthesis of existing views, interpretations can be given a different value, and consequently, the individual can draw conclusions that will move him/her to act accordingly until a new perspective is encountered (Mezirow 2003). Self-reflection can lead to significant personal transformations, and this can be achieved by methods that involve raising consciousness and participation in social action (Mezirow 1997).

In addition to self-reflection, it is necessary to consider the construction of dialogic communities of learning among adults as a way to achieve an information society for all. Learning Communities are defined as a project of social and cultural transformation of an educational place (Elboj et al. 2002; cited by Molina and Ríos 2010), and are based on dialogic learning. According to Molina and Ríos (2010), learning communities emerge as a model that contributes to the inclusion of students with disabilities. The premise of this model, based on dialogic learning, is to achieve a better learning and coexistence among all their members, regardless of their particular characteristics. Essentially, learning communities are a response to diversity in schools (Ainscow 1994).

This paper deals with the design and proposal of methodological strategies to foster a gradual transformation in teachers towards embracing full inclusive education of students with visual impairment in their classroom. This proposal is grounded in dialogic learning, thus leading towards a process of teacher transformation by critical reflection and self-awareness of their current practices so that they move towards improving their teaching practice for diverse students.

The research

The teacher-training programme developed as a didactic proposal was designed by scientists, science educators, and a psychologist, based upon the principles of Universal Design for Learning (UDL) (UNESCO 2015, 20) and dialogic learning. We position UDL as a framework for reducing education barriers, and therefore, providing equal opportunities for all students. Thus, activities aimed to promote a reflective dialogue on the essentials, requirements, and benefits of an inclusive education for science, and how these principles compare to traditional teaching practices, in order to promote a transformation process towards the creation of inclusive learning environments. Since we used the example of visually impaired students to approach inclusive practices, awareness of visual impairment was necessary. The method we selected was to invite teachers to perform all workshop activities while blindfolded. We are aware that this method could also bring to mind a negative perception of visual impairment, but we guided teachers to focus on the positive aspects of the experience, among them: developing skills for assertive communication during instruction, recognising their abilities to perform the experiments without the use of sight, and to observe with all available senses. We also included basic rules for working with students with visual impairment such as: precision in the use of language, classroom setting, and orientation guidelines. Science hands-on activities were multisensorial (Malone and De Lucchi 1979; Soler-Martí 1999) and included both subject-matter content and inquiry-based science teaching principles (Minner, Levy, and Century 2010).

The entire programme consisted of 40 hours divided into three stages: (a) practical instruction, (b) independent work of participants, and (c) guided work of participants with the workshop facilitators. The first stage, practical instruction (12 h), was designed to cover four main aspects: principles and practice of inclusive education, awareness of visual impairment, pedagogical approaches for inclusive education for science, and foundations for the design of inclusive didactic science sequences. This stage set the bases for the dialogic principles of transformation, instrumental dimension, creation of meaning, and solidarity.

Independent work took place after practical instruction. Teams engaged in group/peer dialogue for developing, testing, and implementing sequences of science learning activities in their classroom. They freely selected a science topic from the official curriculum and programme of studies for the level of education they were teaching. Here, the collective constructions of representations, meanings and knowledge acquired an important role. Guided work occurred concurrently with independent work. Facilitators were readily available to guide participants as they reviewed, supported, and provided feedback on didactic sequence proposals for activities prior to their implementation. After a first pilot project and implementation of their didactic sequences, teacher teams elaborated portfolios containing written comments of the process and reflections about the experience of implementing sequences of inclusive science-learning activities in their classroom or with a group of learners of their choice.

The group of participants varied in age (20 to 50 years old), background (80% had post-graduate studies), and work experience, as they were working at different levels including primary school level, special education, high school level or teacher training programmes, and there was one pre-service teacher. Their number of years in service varied from 0 (pre-service) to 24 years; all in-service teachers worked for the public education system and, with one exception, in an urban context. All participants voluntarily enrolled in the workshop, offered in the framework of a state-wide teacher education conference.

In this qualitative study, data were collected from:

- Teachers' portfolios (presented by the team)
- Researchers' observations (non-participative) of classroom activities taking place when teachers implemented their inclusive didactic sequences with their students (12 hours)
- Interviews with participants (semi-structured).

Semi-structured interviews took place five months after the workshop, and were made to one representative of each team with the aim of establishing any changes they had made in their teaching practice. The information gathered from different sources allowed validation by triangulation of the data. To distinguish the origin of the information gathered and shown below, codes are used as follows: I1 indicates interviewed participant and his/her number (I1 for interview of participant from team 1, I3 for interview of participant from team 3, and I4 for interview of participant from team 4), no interview was obtained from team 2. Researchers' notes are coded as RN. P1 refers to the portfolio number 1, P2 to portfolio 2, and P3 to portfolio 3.

Based on the analysis we first identified two modes of implementation of didactic sequences: Mode 1 where teachers merely implemented inclusive science-learning activities with young learners and Mode 2 where teachers included the awareness process prior to the science learning activities (see descriptions below). Next, we identified the emergence of six categories that condense the point of view of the teachers as result of the programme's activities: (a) individuals' awareness as a central axis for the transformation of their practice, (b) emergence of dialogic communities of practice towards teaching students with disabilities, (c) cues of a transformation of knowledge, practices, and context, (d) indication of a need for permanent preparation of teaching for inclusive education, and (e) applicability of the teaching proposal.

Implementation of sequences in classrooms

Four teams of teachers developed proposals of inclusive science-learning sequences. Two proposals targeted topics which are part of the science curriculum (Natural Science) at primary school level: the human digestive system and human nutrition for health. The other two proposals targeted topics on Biology at high school level: one on food and health and another on plant adaptation and nutrition. Examples of the two modes of implementation are presented.

Example of Mode 1 (Team 2)

The team consisted of two female teachers; one teaching at primary school level (A) and another at high school level (B), who developed a sequence on 'Food and Nutrition', a topic belonging to the 3rd grade Natural Science curriculum. The sequence was implemented with a group of 32 students attending a public school in an urban context; in this group, one student had Asperger syndrome, one had a mild learning disability, and five more were described as 'undiagnosed', identifying either a disability or a special need.

Their inclusive didactic sequence contained multisensory activities that started with the identification of food samples which students brought to the classroom for such activities, so they could determine their origin and classify them. Students were asked to freely express their ideas as dialogue was promoted. Next, students were engaged in a game in which they located each food sample within a tactile and colourful representation of the 'Healthy Eating Plate'. Finally, students formed teams to propose examples of a balanced meal and discussed them with their classmates; team work encouraged exchange of ideas and dialogue between peers in order to achieve agreements and prepare their presentations. Both researchers' notes (RN1) and teachers' portfolios describing the experience reported the enthusiasm, motivation, and attention of all students during the activities performed, while providing evidence of an active participation of all students, and particularly noticing the enthusiasm of students with disabilities.

> It is important to notice that all students, including those with special educational needs, presented great motivation during activities; at the moment of making teams, we took special care so they were strategically placed in such a way that there were no more than two students with behaviour problems in the same team. (P2, teacher A)

In their written reflection, both teachers expressed that they would have liked to videotape the students' facial expressions of enjoyment, and mentioned that they were so 'entertained' that did not move from their activities when the recess bell rang. Teachers also reported that students would like all lessons to be similar:

> At the end, they (students) were asked if they would like all classes to be just like that one and responded: "Yeeesss!" (P2, teacher A)

In particular, the high school teacher noted the effect that this experience had on her:

> After learning all this, I am no longer the same; now, in addition to the evaluation, I have in my head what I should do, so that what happened here with these kids can also happen in my class. (P2, teacher A)

It must be noticed that currently in Mexico teachers are undergoing a nation-wide evaluation that will have direct impact on their employment. A highlight of this experience was the interaction and collaborative work between students with disabilities and their classmates, mediated by the science learning activities.

Example of Mode 2 (Team 3)

The team consisted of four teachers (labelled C, D, E, F) working at high-school level. Their sequence was related to the topics of digestion and nutrition. It was carried out during summer recess with four students who were willing to attend school and participate in the activities. Even if there were no specifications of a formal awareness process in the written proposal of the inclusive didactic sequence, interviews and researcher's notes revealed that students performed activities while blindfolded, and teachers' portfolios presented evidence of that.

Activities started with exploration of preconceptions about digestion, the organs involved in the process, and the physical and chemical reactions during it. Next, students carried out experiments to demonstrate digestion using fruits, eggshells, and vinegar. Then, they used three-dimensional representations of the digestive system to identify organs and their function. A final activity included exploration of food samples, grouping them according to their source, and organising them within the idea of a balanced meal. All activities were multisensorial.

Teachers' portfolios reported good student participation in all activities, while becoming aware of how they used other senses to learn:

> Students mentioned that, as they were not able to see, they activated their senses, in particular their hearing; and they also recognised that in this way they can learn too. (P3, teacher D)

In their reflection teachers acknowledged that they could use this methodology in the regular classroom:

> Frequently the methodology that we use, as teachers, is important, given that it is very likely for a student to learn more from the information that his/her body receives and (from) his/her experience, rather than from a teacher's presentation or discourse. (P3, teacher F)

At the end of their reflection, teachers also mentioned their willingness to move towards inclusive practices:

> We are enthusiastic for the work we can do in classrooms, and we are willing to change. (P3, teacher D)

> They talked about the outcome of the activities and stated to be willing to review their teaching practices; they also foresee producing new inclusive activities. (RN2)

Outcomes of the workshop

(a) Teachers' awareness as the central axis for the transformation of their practice

As previously stated, the workshop design included activities to raise awareness and empathy for the condition of visual impairment, supported by reflexive egalitarian dialogue. It emphasised that visually impaired students were recognised as equals, but also as persons with educational needs requiring inclusive education strategies.

> Our design was related to the digestive system and one of the purposes was to raise awareness about visual impairment or difficulties of such disability. But then we also integrated elements related to learning styles and collaborative work. (I1)

> We have ramps but there are other aspects that could help us include students. The school environment has to be prepared; the laboratory and the classroom have to be designed for inclusion. (I4)

As some teachers refer, 'feeling momentarily as blind' helps them experience how a visually impaired student operates in the school context and, therefore, it also raises awareness about students' diversity.

> It also gave teachers the certainty that those students were able to participate in science experimental activities. (RN3)

> I think it makes us become more conscious about how they feel and how we can include them to think. It is not the same to know they can't see rather than living as blind people. Then, precisely, start thinking that through the senses one can develop very interesting practices; although they can't see through a microscope, they can touch and manipulate materials and have an idea of what we are talking about, rather than leaving it to the imagination. (I3)

From those comments, it can be noticed that, triggered by raising awareness, teachers developed a creation for meaning about the condition of visual impairment and the need for including students with such a disability in science-learning processes, under the premise of recognising differences as an opportunity for learning. This created the basis for the construction of dialogic learning communities, which aim not only to integrate students into regular activities, but to include them using teaching strategies to approach topics in a concrete manner.

(b) Emergence of dialogic communities of practice to facilitate inclusive education

From the teachers' interviews, it is possible to observe the construction of communities of practice based on some of the principles of dialogic learning. For example, by egalitarian dialogue in a community of practice, centred on the idea of generating inclusive environments to favour participation of visually impaired students in science learning processes.

> There are teachers who are doing this in other schools; so, we can be in contact with those who are working in inclusion and have them share ideas, experiences and ways to work with students. (I4)

> I work in a kindergarten. We are an integrated school and have a student with visual impairment. So, yesterday we (teachers) met and were developing actions we could do with special needs students. I told them I had done these activities and shared with them how this was, offering to work with them afterschool so they could use these activities with another student. (I1)

We can interpret this was an experience that also raised challenges at their institutional and social levels, as teachers reported they are already promoting the communication of their experiences with colleagues. As they did this with the purpose of generating educational practices that respond to the needs of students with disabilities, we can observe the principle of creation of meaning, which could grow into a broader impact.

It can also be observed that teachers exchanged ideas and experiences, which stimulated a process of reflection that encouraged them to build shared meanings inherent to science teaching and inclusive education.

> … the workshop enhances, gives ideas, to say 'I can do this, I can do these practices. I can take them into consideration', so they can feel as true participants of team work. (I3)

> For example, those who can't see develop enhanced sensitive olfactory skills, more than regular people. We also need to consider safety aspects with him, practices where he can learn but also based on safety. (I4)

(c) Cues of a transformation of knowledge, practices, and context

In parallel to the construction of meaning and significance about disabilities and inclusive education, teachers also appreciated obtaining didactic alternatives, enrichment of ideas, and knowledge, while recognising the need to transform their current version of their teaching practices and the school context.

> When designing activities, we have to take into consideration all elements so he can do it, the entire process or at least a good part of it. Pointing out that he actively participates, so others do not perform the chores that he can do. Let him experiment, live and share his/her experience. (I4)

> Participating in the workshop Science in the inclusive classroom is one of the experiences that most didactic alternatives have given me in my work as teacher. (P1)

Some also identified a compulsory need for transforming their school contexts at different levels to facilitate true inclusion. As for the infrastructure side, modifying laboratories; on the academic side, effective practices or promoting actions concerning the inclusive teaching of science; on the student side, creating or increasing science club members in the case of pre-service teachers.

> The school also has to be prepared, so that laboratory and class environments are designed for them, or the distances, are small things to consider. (I4)

> At the teacher training school, the science club was only open for 7th semester students who chose obtaining their degree with work about science. Then we engaged those kids and others, more to implement the design we did. Implementing was how it was known among students. Today, the club is composed of 35 students of 3rd and 5th semesters who are interested from what they saw. (I1)

> The students, who participated in the implementation of this design, now when they are planning their own work with their students, are using this type of strategies for every topic; that is, using tactile materials. (I1)

As a further consequence of the science club, one of the students is currently preparing his final work for teaching credentials under the modality of teaching practices from the perspective of inclusive education.

> When we implemented it, we did it so they could observe how they could do this with their own students, because we implemented it with 3rd and 5th semester of the bachelor's degree (pre-service teachers), and one of them, who collaborated in the team, is now in 7th semester and he recovered this design of the digestive system to use with his 4th grade students; and with this work he will do a practice report to obtain his degree. (I1)

Data presented above illustrate the participation of one teacher and her student from a teacher credential programme and exemplify the effects a workshop can make in transforming educational practices to promote inclusive education. In other words, her own practice is already having an effect on different levels, directly with her pre-service teachers and then, as consequence, their future students.

(d) Indication of a need for preparation of teaching for inclusive education

The elements of transformation evidenced in previous statements, point to a need for teacher preparation for the practice of inclusive education as part of the curriculum in teacher training schools. Participants also explicitly stated that they need to learn about inclusive education and about design of teaching strategies for students with visual impairments.

> If there was another inclusive workshop, focused on this area, then we'll take it too, keep on preparing ourselves in these guidelines. Keeping in touch with teachers, who are also working with inclusion, and contribute with new ideas, with their experience; teachers who share ways of working with students. (I4)

This suggests that close support with a specialist would be desirable so they can construct dialogic learning communities in order to further improve their teaching practices and generate inclusive environments.

> I believe that we have the need, or it is necessary, to have more courses in this regard. If this kind of practices were to continue, it would be excellent, because we could keep up with those abilities we are developing, improving. (I3)

Here, the idea of creation of dialogic learning communities, based on specialised coaching processes, is positively viewed.

(e) Applicability of the teaching proposal

The implementation of didactic sequences for inclusive education of science allowed significant accomplishments according to the participants; for example, triggering an interest and taste for science among students, independently of the education level, by incorporating tactile materials and separate from old teaching methods based on the transmission of knowledge.

> When they (students) are really doing, touching, it is more attractive for them and more didactic; learning can be better attained, I feel. Sometimes we are tied up too much with teaching theory and they do not understand, they memorise, but do not really comprehend. Hence, in a more practical way, well, simply they are looking at concepts. (I3)

Another important achievement is that, when using the perspective of inclusive education, teachers incorporated UDL principles into their regular classrooms, noticing they can better capture student's attention. This also has implications for teachers' motivation to propose more practical activities that will contribute to a better teaching of science.

> It happened it was recess time and children did not go out, even when [teacher] told them that it was recess time they only looked at her and did not move. They kept doing the work as they were quite entertained, until someone told them to go out and that they could continue when they were back. When they returned, they just continued with no further instructions. (P2)

> With students from the teacher's school (pre-service teachers) I have seen motivation; they are more attracted to work with science and the design of strategies that involves working with senses. Yes, they like it better. (I1)

> I saw them very interested; they said they liked it because it was different to what we normally do. (I3)

Implications for teacher education courses

In the design of programmes for the promotion of dialogue and collaboration among teachers, aiming towards a transformation of traditional teaching practices towards new practices that incorporate elements for inclusion, we used the example of students with visual impairment as a paradigmatic example of how to promote an inclusive education for science. Here, we noticed that the awareness/sensitisation step allowed teachers to be aware of the condition of visual impairment, but more importantly, it seemed to offer elements for new significance of this condition. The fact that teachers implemented their didactic sequences using the same strategy for sensitisation as we used with them, could reveal higher level of appropriation of meaning and greater feeling for people with disabilities. It has to be understood that this was only meant to be an initial strategy, but data indicated that the experience took over and expanded to redefine their role in teaching science for all students.

It was also possible to observe that teachers referred to the importance and need to design inclusive learning environments, but also recognised the necessity to make more profound changes to the fabric of their schools and the school system. This as a base to promote a transformation originated from the practice and not from the political discourse as it is usually presented, and was not implemented in reality. We consider that this level of consciousness manifested by teachers (from interviews, portfolios, and observations) is a first level of change derived from the awareness, and strengthened through the dialogue with their peers. This occurred during the development of their didactic sequences, and further socialisation and reflection of their experience. Designing inclusive didactic sequences was considered also as a new way to work that strengthens their teaching practice.

Participants of the workshops repeatedly expressed the need to implement similar workshops in their schools, because these are not usually available in regular teacher training opportunities nationwide. This is because, in general, they do not feel sufficiently prepared for teaching students with disabilities. Some teachers even suggested that these courses should be mandatory in the teacher training curriculum, given the relevance of acquiring skills and knowledge to generate changes in science teaching from a global inclusive perspective, and because this is missing in the current programmes. This research provides initial evidence to suggest the need for incorporation of innovations at the curricular and practical level for teacher training schools and for in-service teacher development programmes in México and elsewhere.

Final reflections

We have presented evidence that it is possible to stimulate a gradual transformation of teaching practices towards inclusion through a teacher-training proposal that promotes self-reflection based on the dialogic learning principles. The nature of the educational experience we report here resulted in the generation of learning communities engaging in collective construction of inclusive teaching practices. Taking into account the principles of dialogic learning that could be used to illuminate our findings, we can summarise that:

(1) Awareness of the needs of students with visual impairments, and the collaborative work to design inclusive didactic sequences, were key elements that induced a reflective and egalitarian dialogue between teacher peers about the importance of creating inclusive environments for science learning.
(2) We observed the instrumental dimension and solidarity principles as teachers were willing to modify their teaching approaches to offer better learning opportunities to all students.

(3) Participants who promoted changes in their schools towards inclusive education of science emphasised the principle of creation of meaning.
(4) In the example of the teachers who worked with the elementary school group, it was noticeable that the principle of equality of differences, when they observed that activities designed for visually impaired students were also more engaging for students with other disabilities.

We can think of Mezirow's reflective judgement as the 8th principle of dialogic learning, evident in teachers' reflections about their experience of participating in the workshop programme and how this experience helped them envision transforming their teaching practices. Gradual changes can induce greater changes in the learner, such as a paradigm change, which aligns with the concept of transformative learning as the learners become self-reflexive and move towards a frame of reference that is more inclusive and socially responsible (Mezirow 1997). This also coincides with other authors (see Calleja 2014) who conceptualise adult transformative learning as a 'structured shift experienced by individuals who allow themselves to learn from their own experience within a community of practice' (2014, 117).

The case we are illustrating here relates more broadly to a discussion on how to promote inclusive education and the creation of engaging learning environments to benefit all students. Based on the outcomes, we suggest that, through experiences like the one we offered, is possible to gradually transform the predominant science teaching practices in the regular classroom, whilst opening the opportunity for inclusive practices, developed under a dynamic of new meanings and valuing of the diversity of students. To attain this goal, the principles of UDL are meant to guide educators to make the curriculum accessible to students with different abilities, learning styles and within diverse contexts. Thus, to generate inclusive schools and inclusive science classrooms, it is desirable that all teachers develop communities of practice to promote the principles of UDL, so they may design teaching/learning activities for all students towards the ultimate goal of reducing educational inequalities. In conclusion, we find evidence supporting the need for teacher education programmes to employ dialogic approaches with pre-service and in-service teachers that align with the most recent educational model in México.

Acknowledgements

Thanks to the University of Sonora (Universidad de Sonora) for the facilities given to offer the workshop. We are indebted to all teachers who participated in the workshop and are now active agents of change towards making inclusive education for science a reality. We thank Luis R. Sandoval for editing assistance.

Disclosure statement

No potential conflict of interest was reported by the authors.

Funding

This work was supported by the SEP-CONACYT Mexico [grant number 0177010]; UIA-FICSAC [grant number F 132014].

ORCID

Cristina G. Reynaga-Peña http://orcid.org/0000-0003-3039-551X
Marisol Sandoval-Ríos http://orcid.org/0000-0002-0007-2309
José Torres-Frías http://orcid.org/0000-0002-0708-2745
Carolina López-Suero http://orcid.org/0000-0003-0131-5760
Adrián Lozano Garza http://orcid.org/0000-0002-2617-0572
Maribel Dessens Félix http://orcid.org/0000-0002-0506-9248
Marcelino González Maitland http://orcid.org/0000-0003-1651-2388
Jorge G. Ibanez http://orcid.org/0000-0003-3247-6751

References

Ainscow, M. 1994. *Special Needs in the Classroom: A Teacher Education Guide*. Paris: UNESCO.

Atwood, R. K., and B. R. Oldham. 1985. "Teachers' Perceptions of Mainstreaming in an Inquiry Oriented Elementary Science Program." *Science Education* 69 (5): 619–624. doi:10.1002/sce.3730690504.

Aubert, A., A. Flecha, C. García, R. Flecha, and S. Racionero. 2008. *Aprendizaje Dialógico en la Sociedad de la Información* [Dialogical Learning in the Information Society]. Barcelona: Hipatia Editorial.

Booth, T., and M. Ainscow. 2002. *Index for Inclusion-developing Learning Participation in Schools*. Bristol: Centre for Studies on Inclusive Education, CSIE.

Calleja, C. 2014. "Jack Mezirow's Conceptualisation of Adult Transformative Learning: A Review." *Journal of Adult and Continuing Education* 20 (1): 117–136. doi:10.7227/JACE.20.1.8.

Cawley, J. F. 1994. "Science for Students with Disabilities." *Remedial and Special Education* 15: 67–71.

Elboj, C., I. Puigdellívol, M. Soler, and R. Valls. 2002. *Comunidades de Aprendizaje: Transformar la Educación* [Learning Communities: Transforming Education]. Barcelona: Graó.

European Agency for Development in Special Needs Education. 2010. *Teacher Education for Inclusion – International Literature Review*. Odense: European Agency for Development in Special Needs Education.

Flecha, R. 1997. *Compartiendo Palabras: El Aprendizaje de las Personas Adultas a Través del Diálogo* [Sharing Words: Adult Learning Through Dialogue]. Barcelona: Paidós.

Flecha, R., J. Gómez, and L. Puigvert. 2001. *Teoría Sociológica Contemporánea* [Contemporary Sociological Theory]. Barcelona: Paidós.

Flores-Camacho, F. 2012. *La Enseñanza de la Ciencia en la Educación Básica en México* [The Teaching of Science in Primary Education in Mexico]. México: Instituto Nacional para la Evaluación de la Educación, INEE.

García, R. 2012. "Out of the Ghetto: Psychological Bases of Dialogic Learning." *International Journal of Educational Psychology* 1 (1): 51–69. doi:10.4471/ijep.2012.04.

Jones, M. G., J. Minogue, T. Oppewal, M. Cook, and B. Broadwell. 2006. "Visualizing Without Vision at the Microscale: Students with Visual Impairments Explore Cells with Touch." *Journal of Science Education and Technology* 15 (5–6): 345–351. doi:10.1007/s10956-006-9022-6.

Malone, L., and L. De Lucchi. 1979. "Life Science for Visually Impaired Students." *Science and Children* 16 (5): 29–31.

Mastropieri, M. A., and T. E. Scruggs. 1992. "Science for Students with Disabilities." *Review of Educational Research, Winter* 62 (4): 377–411.

McGinnis, J. R., and G. P. Stefanich. 2006. "Special Needs and Talents in Science Learning." In *Handbook of Research in Science Education*, edited by S. K. Abell and N. G. Lederman, 287–318. Mahwah, NJ: Lawrence Erlbaum Press.

Mezirow, J. 1997. "Transformative Learning: Theory to Practice." *New Directions for Adult and Continuing Education* 74: 5–12.

Mezirow, J. 2003. "Transformative Learning as Discourse." *Journal of Transformative Education* 1 (1): 58–63. doi:10.1177/1541344603252172.

Minner, D. D., A. Levy, and J. Century. 2010. "Inquiry-based Science Instruction – What is It and Does It Matter? Results from a Research Synthesis Years 1984 to 2002." *Journal of Research in Science Teaching* 47 (4): 474–496. doi:10.1002/tea.20347.

Molina, S., and O. Ríos. 2010. "Including Students with Disabilities in Learning Communities." *Psychology, Society, and Education* 2 (1): 1–9.

NAP (The National Academies Press). 2015. *Next Generation Science Standards: For States, by States. NGSS Lead States*. Washington, DC: The National Academies Press.

Norman, K., D. Caseau, and G. P. Stefanich. 1998. "Teaching Students with Disabilities in Inclusive Science Classrooms: Survey Results." *Science Education* 82 (2): 127–146.

Prieto, O., and E. Duque. 2009. "El Aprendizaje Dialógico y sus Aportaciones a la Teoría de la Educación." [Dialogic Learning and its Contributions to Education Theory.] *Education in the Knowledge Society* 10 (3): 7–30.

Ruiz, G. 2012. "La Reforma Integral de la Educación Básica en México (RIEB) en la Educación Primaria: Desafíos para la Formación Docente." [The Comprehensive Reform of Basic Education in Mexico (RIEB) in Primary Education: Challenges for Teacher Education.] *Revista Electrónica Interuniversitaria De Formación Del Profesorado* 15 (1): 51–60.

SEP (Secretaría de Educación Pública). 2012. *Atención Educativa para la Inclusión* [Educational Care for Inclusion]. Plan de Estudios 2012. México: Secretaría de Educación Pública, SEP.

SEP (Secretaría de Educación Pública). 2017. *Modelo Educativo para la Educación Obligatoria* [Educational Model for Compulsary Education]. México: Secretaría de Educación Pública, SEP.

Shulman, Lee 1987. "Knowledge and Teaching: Foundations of the New Reform." *Harvard Educational Review* 57 (1): 1–23.

Soler-Martí M. A. 1999. *Didáctica Multisensorial de las Ciencias: Un Nuevo Método para Alumnos Ciegos, Deficientes Visuales, y También sin Problemas de Visión* [Multisensory Science Teaching: A New Method for Blind, Visually Impaired and also for Students without Visual Problems]. Barcelona: Paidós.

UNESCO (United Nations Educational, Scientific and Cultural Organization). 2009. *Policy Guidelines on Inclusion in Education*. Paris: UNESCO.

UNESCO (United Nations Educational, Scientific and Cultural Organization). 2015. *Embracing Diversity: Toolkit for Creating Inclusive, Learning-friendly Environments. Specialized Booklet 3: Teaching Children with Disabilities in Inclusive Settings*. Paris: UNESCO.

Villegas-Reimers, E. 2003. *Teacher Professional Development: An International Review of the Literature*. Paris: UNESCO International Institute for Educational Planning.

4 STEM outreach activities

An approach to teachers' professional development

Farzana Aslam, Arinola Adefila and Yamuna Bagiya[§] (England)

ABSTRACT

STEM outreach programmes in secondary schools are mediated by STEM teachers who are responsible for organising, implementing and evaluating the activities with a view to promoting STEM subjects. However, research investigating teachers' STEM roles and professional development through participation in outreach activities is limited. This paper explores teachers' views of STEM activities, how they understand their role as primary facilitators and the impact of their STEM engagement on their professional development. STEM outreach provides distinct opportunities for situated and dialogic learning. STEM teachers' effectiveness in engaging students by demonstrating the importance and relevance of STEM subjects in the real world and engaging captivating STEM outreach partners not only supports students learning but also enhances their opportunities to develop their knowledge and skills. Semi structured interviews were conducted with eleven teachers from six different schools in the West Midlands, United Kingdom. The data reveals STEM facilitators become the 'face' of STEM to their pupils. The expertise, knowledge and partnerships STEM facilitators develop, broadens their capacity to deliver teaching imbued with real world applications and improve students' understanding of the range of new and upcoming careers available. Most importantly, participating in STEM outreach activities help teachers maintain and reinforce their own sense of identity as STEM professionals. Outreach activities provide opportunities for teachers to interact with leading scientists and obtain exposure to cutting-edge research.

Introduction

A contemporary focus on Science, Technology, Engineering and Mathematics (STEM) subjects and careers has led to a proliferation of outreach activities for students of all age groups in UK and globally (Vennix et al. 2017). Outreach includes interactive STEM workshops, STEM ambassador presentations, mentoring schemes, STEM placements, visits to industry, master classes, competitions and consultations about STEM careers, activities geared towards

[§]Present address: Maths Learning Centre, De Montfort University, Leicester, UK.

making STEM subjects fascinating and appealing. STEM outreach practitioners seek to enhance and enrich students' learning experience and expose them to areas of STEM that may not be delivered as part of the regular curriculum (Laursen et al. 2007; Hargreaves 2010). Practitioners represent professional institutions, voluntary organisations or universities and provide crucial opportunities for students to engage in real world STEM challenges that have the potential to stimulate interest in STEM careers (Kesidou and Koppal 2004).

Outreach programmes are embedded in learning environments that should demonstrate the relevance of STEM and inspire, intrigue and motivate students to study STEM subjects beyond compulsory schooling (Turner et al. 2007). However, for STEM outreach to be efficient and successful, the STEM teachers who provide the bulk of STEM teaching need to work collaboratively and strategically (National Audit Office 2010) with practitioners to ensure the right students are selected for programmes and the activities dovetail with continuing teaching and projects. Furthermore, teachers' involvement in outreach could promote their own professional development (Royal Academy of Engineering 2013). As teachers are the vanguards of the school-university/industry partnerships, they formulate through outreach initiatives a powerful space for advancing their own continued professional development (Maistry 2008). The STEM teachers in effect are able to develop distinct communities of practice (CoP) (Lave and Wenger 1991) which serve as engines of learning and instruments for personal development.

Unfortunately, students' access to STEM activities are heavily dependent on the STEM teachers in a particular school and the school management (National Audit Office 2010; Straw and Macleod 2015). Practitioners reveal that they often engage with the same set of schools, corroborating the notion that proactive teachers employ their services more regularly (Holman and Finegold 2010; Murphy and Machin 2011). STEMNET, the National STEM Learning Centre and Network, responsible for coordinating outreach activities across the UK, acknowledges the need to provide more support for teachers to enable them set up STEM clubs and involve practitioners (Straw and Macleod 2015). However, teachers do not necessarily have the competencies, opportunities, motivation or enthusiasm to develop and sustain STEM outreach (Baker-Doyle and Yoon 2011). Straw and Macleod (2015) suggests that teachers do not always have the capabilities or administrative support to efficiently provide outreach services. Additionally, because STEM activities can be viewed as detached from more pressing curriculum goals, teachers are not encouraged or equipped by the educational system or their school management to be efficient STEM facilitators. Each individual school's commitment to STEM education will be influenced by the competencies, and motivation of the STEM teachers, attitudes of head teachers and STEM department coordinators. These factors account for the significant variation in STEM outreach engagement both locally and regionally and this may be the reason why some schools do not engage with STEM practitioners at all (National Audit Office 2010). Furthermore Bybee (2010) argues that although there is a lot of commitment to establishing STEM activities, the policy initiatives that will enable teachers to do this is completely lacking. The infrastructure, resources and training to enable teachers to be efficient STEM facilitators the system needs seems to be absent.

This paper presents research which explores teachers' professional development through outreach activities and underlines the need to reform teacher education and implement a framework which enables teachers to successfully engage with outreach activities thereby developing their role as STEM facilitators. Teachers' crucial role in coordinating and facilitating outreach events can be pivotal for a school's STEM education. The attitudes and enthusiasm of a teacher often influence the direction and intensity of STEM engagement. This has a direct correlation on the quality of experience gained by the students through outreach participation (Murphy and Machin 2011; Laursen, Thiry, and Liston 2012). Key themes the research focuses on are: teachers' perspectives on engaging with STEM outreach, how the outreach programme as a whole affects their understanding of a teacher's facilitation role and professional development, and the support and education they require to effectively engage with outreach. The data presented in the paper is gathered from in-depth qualitative semi-structured interviews with secondary school/college teachers involved in facilitating STEM outreach activities.

Role of STEM outreach in supporting the learning of teachers

STEM outreach focuses on informal activities (Mann and Oldknow 2012). The aim of STEM outreach strategy is to inspire students and enable them to appreciate STEM careers. STEMNET an education charity in the UK (Straw and Macleod 2015), for example, encourages teachers to engage with ambassadors and practitioners so that their students can meet inspiring role models who work in STEM careers and participate in real life projects and workshops which will directly promote students' understanding of real world applications of STEM subjects. STEMNET provides some training and Continuous Professional Development (CPD) for teachers, particularly with respect to how to effectively utilise the resources it supplies to schools. STEMNET also coordinates a number of STEM outreach activities, providing the teachers with access to STEM practitioners and the support they need to set up STEM clubs. The STEMNET support and provision presupposes that teachers have the capabilities and skills required to run the STEM clubs. Teacher education is predominantly focused on the delivery of curriculum (Vennix, Brok, and Taconis 2017). The extra-curricular activities and innovative practices, resources and activities offered by STEMNET often can only be delivered by teachers who have enhanced skills, particularly the capability to network with others, manage and coordinate activities across the school and engage with the financial, social and structural issues involved in being a facilitator. Many teachers will find this a challenging experience (Baker-Doyle and Yoon 2011).

STEM activities can provide experience of hands-on STEM projects that motivate, inspire and bring learning and career opportunities to life (Straw and Macleod 2015). Teachers are expected to spearhead STEM clubs and act as coordinators of STEM outreach but there is little training provided for teachers on how to achieve this and engagement with informal outreach in STEM is not embedded in teacher education programmes. There are currently 3,500 STEM clubs across the UK and 71% of schools in a 2015 survey indicated they will run a STEM club in the future (Straw and Macleod 2015). What STEM outreach essentially provides is a great opportunity to situate learning for students, grounding theoretical knowledge in real life situations. STEM practitioners also utilise dialogic learning strategies, supporting students to engage in interactive discourse and enhancing reflexive thinking. These skills are extremely important not only for students hoping for careers in STEM but for supporting the development of vital life skills in modern knowledge economies. STEM outreach therefore provides an opportunity for teachers to learn how to utilise new pedagogies in their practice and provide students with a broader understanding of new research and information relevant to STEM. Traditional teacher education is not sufficient to provide teachers with a comprehensive, up-to-date understanding of how knowledge is being transmitted and reproduced in today's economy. Teachers will need CPD to continue to access STEM and keep in touch with STEM practitioners.

Practitioners and teachers working together to deliver STEM outreach promote a 'collaborative learning community' thus fostering the principles of appreciative inquiry (AI) (Cooperrider and Srivastva 1987, 129). This approach repositions the focus from mere problem identification to an investigation of the range of solutions possible; students and teachers working together to understand the solutions to problems in real time (Walker and Salt 2006). This kind of pedagogic space is often informal and fosters open, honest discussions between learners and teachers. The egalitarian dialogue can enhance effectiveness and integrity amongst the students, teachers and the practitioners (Cooperrider and Srivastva 1987). Teachers are the means through which the dialogic, situational pedagogy of the STEM outreach activities can be corroborated in classroom teaching and this provides interactions with the wider learning environment. The aim of this study is to explore teachers' role in STEM outreach and how this influence their professional development.

Methodology and data collection

The primary aim of the study was to obtain specific in-depth information from those who are able to provide it rather than information which would generalise and be representative of the wider population (Creswell 2007). A phenomenological approach was adopted and this involved 'describing the lived experience' (Cohen, Manion, and Morrison 2011) of the participants, including their everyday practice, activities and decision-making processes. The study aimed to understand teachers' subjective experience of STEM outreach (Patton 1990). A phenomenological approach was adopted for the study as it enables an exploration of teachers' wealth of knowledge, expertise in this field and perspectives on the impact of STEM outreach.

Semi-structured interviews were used to gather data from teachers across different state schools in the West Midlands, UK. A protocol was developed for the study based on the case study method of Yin (1984). The themes explored in the protocol were focused on the interaction between STEM practitioners and the learning environments. It was important for participants to be comfortable expressing their own point of view to the independent researcher, unconnected with the schools' management or STEM networks.

Purposive sampling was used to ensure that a diverse population of teachers involved in STEM outreach were interviewed (Cohen, Manion, and Morrison 2011). This involved ensuring STEM teachers with a varied range of experience and responsibilities were recruited. The sample included teachers who specialised in all STEM subjects: biology, chemistry, physics and mathematics. Details of their role as a facilitator varied across the sample. Teacher 5 was a support teacher and Teacher 8a/8b represented two teachers from the same school who were interviewed together. It was also important to ensure our sample was recruited from diverse state schools. We used the Free School Meal (FSM) categorisation to achieve this, recruiting from categories 2 – 4 as shown in Table 1 below. The aim of our purposive sampling was to include teachers who taught students from a range of economic backgrounds. Students from low socio-economic backgrounds who are more likely to receive FSM are known to have limited out-of-school engagement with STEM careers and activities.

Table 1. An overview of the teacher participants in the study.

Teacher	Specialist subject	Free meal category	No. of years of teaching	School year they teach	Role title
1	Biology	4	–	Years 7–13	Head of Science
2	Biology	2	–	Years 7–13	Associate Head Teacher
3	Biology	2	6 years	Years 7–13	Acting Head of Science STEM and Environmental club coordinator
4	Biology	4	–	Years 7–11	Subject Specialist
5	Biology	2	1 year	–	Intervention manager in the science departmentSTEM organiser
6	Chemistry	3	23 years	Years 7–13	Subject Specialist
7	Chemistry	2	15 years	Years 7–13	Associate Assistant Head Teacher Advanced skills teacher
8a/8b	Physics	4	10 years	Years 7–13	Subject Specialist
9	Physics	3	7 years	Years 7–13	Subject Specialist
10	Physics	4	12 years	Years 7–13	Lead practitioner
11	Mathematics	2	–	Years7–13	Head of Maths

The questions of the semi-structured interview focussed on exploring the insights in learning environments created by STEM Outreach and factors that influence decision-making when organising STEM outreach activities. The teachers were invited by email to participate in the study. Interviews took place at the teachers' school or at a nearby café as requested by the interviewees. Data was transcribed and thematically analysed using NVivo. The researchers independently reviewed the themes through the phases of the analysis. All data was confidential and teachers cannot be identified.

This research has followed the professional code of practice highlighted by the British Educational Research Association (BERA) Ethical Guidelines for Educational Research (BERA 2011). It has also followed Coventry University's ethical guidelines and ethical approval was gained before conducting this research. In addition, before the start of the interviews, oral and written permission was obtained from the interviewees to record the interviews. They were informed of their participation being voluntary and were reassured about the confidentiality of their responses throughout this study. Careful consideration was also given towards protecting the participants' identity.

Findings

The findings of the study highlighted the teachers' significant role as STEM outreach facilitators, the importance of the Community of Practice (CoP) they draw on for support and information, teachers' professional development through STEM outreach and the challenges the teachers' face delivering and participating in STEM outreach.

Teacher's role and identity as STEM outreach facilitators

The findings suggest that as part of their need to collaborate with practitioners from industry and academia to organise STEM outreach activities, teachers need to develop a dual role and skills as STEM facilitators. Practitioners and teachers working together to deliver STEM outreach promote a collaborative learning community. The STEM activities centre on using the principles of appreciative inquiry (Cooperrider and Srivastva 1987), supporting both formal and informal learning through dialogic interaction. This provides a unique opportunity for teachers to learn from those in industry and engage with research scholars. One teacher explained how a STEM ambassador challenged him to engage students creatively using dialogic interaction.

> That person came in with nothing. He didn't even use my blackboard. He didn't use my PowerPoint, he didn't use my computer; he was just standing there, but he enthused the students so much with his talk, and the way he gave that talk, that was brilliant. So you don't need resources to engage students and get them interested
>
> Teacher 6

In effect the teachers are developing their teaching methods and learning because of their roles as facilitators. STEM facilitators appreciate the value of conceptualising learning as transactional and transformative and this is an integral part of STEM outreach. First, it shapes the identity (Bandura 1986) of the teacher acting as STEM facilitator, casting them as a member of the scientific community. The teacher therefore networks with practitioners and ambassadors from the STEM community to develop their understanding of STEM. Bandura (1986) suggests that social identities influence the production of both social and cultural norms as well as self-efficacy and agency. Caza and Creary (2016, 259) argue that 'professional identity is an important cognitive mechanism that affects workers' attitudes, affect and behaviour in work settings and beyond'. The STEM facilitators routinely discuss how their role shapes their behaviours and actions. This is particularly evident in their teaching and delivery of STEM subjects but also in their determination to provide STEM club opportunities for their students, even when they have to make sacrifices of time and effort. Some teachers see their facilitator role as a duty.

> I think, what's the point in being a teacher unless, actually, you are trying to have a positive impact on the life choices that they make? So not me as a teacher; I see my role as … a facilitator to show them choices.
>
> Teacher 7

> It's just organising, a lot of money and a lot of time, really. Organising the different trips and making sure that they go to what's out there you know, what's available out there. Trying to mop up as there's lots of things out there that they can do but we can't do everything, because we haven't got the money or the time to do it.
>
> Teacher 3

Second, outreach practitioners and ambassadors working collegially with the teacher, seek to inspire students to join the community of scientists, engineers and technology experts who help transform the real world, creating the medicines, gadgets and research that change how we live. As such, beyond the skills and knowledge shared at an outreach event, the positioning of the STEM teacher and the relational linkages between STEM subjects and the everyday technological products and innovations are accentuated. This supports the acculturalisation of the students as well as the teachers (Lave and Wenger1991). STEM outreach that is decoupled from a transactional, transformative STEM classroom pedagogy will be weak and ineffective because it undermines the agency of students to continue to co-construct and develop their learning of STEM alongside the teacher within the school's learning environments. Once again, the centrality of the teacher as a co-learner is emphasised n pedagogic spaces where learning is interactive. This leads to the second important finding, which reveals the importance of STEM teachers benefitting from CPD in CoPs.

Teachers' in STEM related CoP

The findings from our study suggests that teachers develop tremendously from their interactions with the communities and resources that enable them to provide STEM activities and clubs. STEMNET and other charities provide specific training and resources to support teachers' CPD so they can use STEMNET resources. However, beyond this kind of formal training, networking with other STEM facilitators, STEM ambassadors and practitioners significantly provides advantages for teachers to develop professionally. This learning is increasingly viewed as a social process (Vygotsky 1978; Bakhtin 1981). Modern pedagogies are often embedded in social learning environments to enable learners to improve their understanding of abstract concepts (Koschmann 1999). Lave and Wenger (1991) advance the notion of collaborative learning in her seminal work about how social interaction contextualises learning. They describe learning 'as an integral part of generative social practice in the lived-in world' (1991, 35). Lave describes how CoPs are utilised to construct meaning and consolidate knowledge (Lave 1986); learning objects help to mediate this meaning. STEM outreach draws extensively upon the ideas of collaborative learning by situating learning in settings of collaboration (Bakhtin 1981; Lave and Wenger 1991). Besides this opportunity for transformative learning through interactive dialogue, Bakhtin's (1981) concept of dialogism suggests a relational link between language, interaction, and social transformation. For Bakhtin, meaning is created through processes of reflection between people (Wegerif 2011). Bakhtin advances our understanding of learning as a dynamic and process-based activity (Koschmann 1999) rather than a discrete exercise taking place in the mind of one individual learner.

> Well ... usually a lot of things happen by word of mouth. You know, sort of, 'Well, that was a good thing. I can pass it on to various other people.' As far as the STEM network goes, STEM network works very well.
>
> Teacher 6

> I think the STEM network is particularly good at, you know, seeking out teachers. But, yeah, I've contacted people through STEM and through university contacts.
>
> Teacher 9

> We do get STEM coordinators and the actual hub of distribution is brilliant and they do send out a lot; it is almost that we get too much and it's finding out which one would be best for your groups.
>
> Teacher 8

The study showed that the opportunities STEM facilitators gained from engaging with STEM networks and organisations were different from any benefits they would have gained from being part of a department, committee or project teams. CoPs share three specific domains (Wenger, McDermott, and Snyder 2002).

(1) Knowledge: a common body of knowledge within the community.
(2) Community: commitment to forming a group for networking.
(3) Shared practice: sharing of ideas, resources and strategies.

The teachers are expanding, creating and sharing their knowledge to develop their individual capabilities. The teachers discuss how because of their networking and interactions there was an emergence of shared practice. Teachers and practitioners exchanging ideas about good practice and how best to support the students into STEM related degree programmes at university.

> You know, anything that was long-term, to me is more valuable. The long-term projects because they're always doing it, they're always thinking about it. It's there at the front of their head ... if they don't see something at the end of it, the numbers will dwindle, even if they think it is fun.
>
> Teacher 10

This teacher is reflecting on what she has learned from sharing resources and collecting practices with the STEM ambassadors and practitioners. A CoP is unique because it evolves to support a shared goal (Wenger, McDermott, and Snyder 2002), in this case, engaging students in STEM activities so they can choose STEM careers.

Teachers can be very influential in shaping a student's attitude towards a subject and their career choices (Maychell et al. 1998;). Students' attitudes to, perception of and understanding of, STEM subjects and careers are formed and developed at different stages during their educational journey (Hargreaves 2010; Adecco Group UK and Ireland 2015). Therefore, it is important that the STEM facilitators are supportive and capable through the students' educational journey to inform and inspire them. STEM ambassadors and practitioners only see students intermittently, but teachers introduce them to the STEM CoP. These STEM CoPs enable teachers to acquire most relevant and up-to-date information about STEM careers so they can advise students appropriately. Research (CEI 2007) has shown that teachers do not feel well-informed about science careers and have few opportunities to update their own careers awareness through courses or placements. STEM outreach can provide these opportunities for professional development of teachers. The teachers also highlight other ways through which their engagement as STEM facilitators improves their opportunities for CPD.

Teacher professional development through STEM outreach

The dynamic nature and innovations in STEM have made teachers' professional development a most important priority in the UK and Globally (Baker-Doyle and Yoon 2011). The complex and multidisciplinary nature of STEM subjects have made teachers' professional development a challenging task. One emerging approach to the creation of successful professional development programmes is to support the development of teacher networks, both formal and informal (Baker-Doyle and Yoon 2011).

This study indicates that in STEM outreach teachers and practitioners work collegially to inspire students to engage with STEM careers. As such, beyond the skills and knowledge shared at an outreach event, the positioning of the STEM teacher and the relational linkages between STEM subjects and the everyday technological products and innovations are accentuated. STEM outreach gives access to interesting and exciting, contemporary research information in a form which teachers can use in their class to innovatively deliver the national curriculum.

In the study teachers highlighted how participating in STEM outreach activities has directly contributed to their own understanding and benefited them personally. For instance, it has provided one teacher with a stress-free environment where she can interact and engage with her students without worrying about an exam.

> When you've got the students in the classroom, you've got a target, you've got a focus that you're under pressure to obviously cover a part of the curriculum, you've got exams in the distance in your mind, you know, you've got to be focused on that. When you do extracurricular, we can do whatever we want, it's a more relaxed atmosphere, you get time to get to know the students in a more relaxed way, they're able to come up with their own ideas. It's fun for them because they're doing what they want to do... that's probably the best thing, really spending time with the students and not having the pressure of having to get them ready for an exam or get a piece of coursework done or you know, you've got a test coming and you can't afford to waste time.
>
> Teacher 3

Teacher 4 indicated how collaboratively working with outreach practitioners can result in effective learning environments:

> I go to meet them and then we plan sessions together and then they'll come and say 'look it's only because I have done a lot of work with them so they know me so they're quite happy for me to go in and co-plan' and they'll give me ideas that will say they've got this and this: what do you think would work; yes so we co-plan and that's when it's really effective.
>
> Teacher 4

Teacher 7 indicated that getting involved in STEM club was very beneficial for new members of staff:

> And for the last four, nearly five years, we have had STEM club every week on a Thursday. It's been quite beneficial for a lot of the younger members of staff in science, ones that have come into teaching. They've had a [chance to get] a lot of experience. They've helped out with the extracurricular club and it has also meant that we were able to run the two clubs, as well.
>
> Teacher 7

The teachers reveal how STEM activities ultimately support their personal development and they use the opportunities to create a pedagogic space for themselves, to be reflexive about their own practice, to plan and co-plan with practitioners. They are able to learn with students collaboratively in a way that is transactional and transformative for themselves as well as the students. Nevertheless, STEM outreach presents challenges for teachers' professional identity as well as, socially and culturally.

Challenges of providing effective STEM outreach

The teachers in the study highlighted four key challenges in providing effective outreach which included administrative support, funding, time and recognition. In describing the problems developing STEM outreach, Teacher 6 expressed frustration that the students could not be provided with more opportunities.

> If we would want something like this on a bigger scale, it would be easier if the school gets involved and it is done in the school time, in an assembly or the whole of year 8s, for example, are targeted.
>
> Teacher 6

Though the teacher is highlighting the need for more administrative support to organise the STEM outreach for more students, this is also acknowledging an apparent contradiction in the STEM outreach process, as a STEM teacher they are taking on the full commitment of running STEM outreach. The school management is not engaging with the STEM outreach process as it ought to. In these types of ambiguous circumstances (Niemi 1997), teachers begin to question their roles and responsibilities. In some cases, the teachers were burdened by playing the multiple roles of teacher and STEM facilitator.

> Everything which I bring in for the STEM club, up until now, was being resourced by myself. If I was using some chemical, if I was using some material, it all came from me. The school has to do more if they want this.
>
> Teacher 6

On the other hand, Teacher 10 was able to negotiate the multiple roles and identities seamlessly, mainly because there was support and funding from the school management. The school acknowledged the importance of outreach events which are embedded into the school's curriculum and most importantly its ethos:

> We do have also in this school four enrichment enterprise days, where the timetable is suspended and then different year groups do different things. So the school actually does good in that, in that they collapse the timetable, or they suspend it, so these things can happen.
>
> Teacher 10

Teachers with high level of commitment towards STEM outreach tried to offer as many activities as possible regardless of the support they had. These teachers seemed to identify the STEM provision as part of their duty and identified more strongly with the STEM community. They sought reliable practitioners, they maintained links with ambassadors and actively engaged with STEM networks. In this regard their STEM facilitator identity is a priority for them. Bleakley (2006) argues that professional identity is continuously being constructed through discourse and interaction with others. This means identities are continually being constructed and altered. This negotiation of professional identity is very challenging for teachers providing STEM outreach with little or no support administratively.

> If you could come up with a legal framework that works for doing that, then and I think very strongly, and the evidence of my department says the same, if you leave teachers to it, they're here because they like conveying science on to students. If you leave them space, then you will get benefit. You will get them doing enrichment.
>
> Teacher 6

There was a common view that in some schools there was a mismatch between the supposed importance of STEM and the outreach programme and the value demonstrated by the resources, time, funding and support allocated to the process. In addition to this, the teachers did not feel they obtained the recognition they deserved for STEM outreach and activities. Teachers wanted to be acknowledged for the work they were doing. There is no suitable structural or policy framework for supporting these types of activities; though the formal and informal CPD opportunities inherent in participating in STEM facilitation are implicit, there could be a more explicit formal acknowledgement of the personal development teachers undertaking STEM facilitation experience.

Discussion and conclusions

A collegial relationship between STEM teachers and practitioners places the teacher in their role as a socialised member of a professional learning community. The teacher and practitioner work together to showcase how their specialised knowledge is shared and develop by interaction. Teachers engaged in STEM facilitation benefit from being in what can be termed STEM communities of practice. In these pedagogical spaces, created by STEM outreach and facilitation, they share and collate practices and resources which improve their development as effective teachers. This kind of CPD goes beyond simply providing STEM as any event, policy, programme or practice that promotes STEM awareness (Bybee 2010) to developing a sustainable STEM network for practitioners, teachers and students to share emerging ideas and information. The STEM networks provide meaningful, practical, active and authentic shared learning experiences (Maistry 2008).

In response, the teacher also enhances the social learning environment in the school, this is crucial to a dialogic or situational learning experience for the students (Lave and Wenger 1991; Laursen et al. 2007). In these contexts, students are able to learn more effectively (Cooperrider and Srivastva 1987) and scaffold their learning with support from teachers and practitioners as co-constructors.

Furthermore, teachers highlighted the need for more recognition, as well as financial and administrative support from schools to enable them effectively support STEM outreach. Their role as teachers is complemented by their second role as STEM facilitators. However, managing these multiple, professional identities can be complex and challenging. The support they need from the school could be formalised within a framework that acknowledges the time, effort and commitment they put into supporting STEM outreach activities, particularly, those engaging in out-of-hours' work. Organised STEM networks also need to formalise the support they provide and demand that appropriate social and financial mechanisms be put in place to support teachers providing STEM outreach. STEM CoPs need to be acknowledged and researched more closely so that we can understand and balance the needs of members and the STEM community as they change with time.

Teachers' involvement in outreach can promote their own professional development. Firstly, they remain engaged with recent developments particularly in relation to university entry requirements and career options available to STEM graduates. Secondly these teachers have opportunities to interact with leading scientists and obtain exposure to cutting-edge research. However, perhaps the most important form of professional development to accrue from teachers participating in STEM outreach is the maintenance and reinforcement of their own sense of identity as a STEM professional. In essence, by participating in the CoPs (Lave and Wenger 1991) and engaging in the STEM activities and social interactions, the teachers are developing and maintaining the networks that transform them as well as their students.

Disclosure statement

No potential conflict of interest was reported by the authors.

ORCID

Farzana Aslam http://orcid.org/0000-0002-3242-6167
Arinola Adefila http://orcid.org/0000-0001-9759-0104
Yamuna Bagiya http://orcid.org/0000-0001-6199-334X

References

Adecco Group UK and Ireland. 2015. *The Gender Agenda: STEMing the Gap*. Accessed December 8, 2017. http://www.adeccogroupuk.co.uk/unlocking-britains-potential/the-gender-agenda.aspx

Baker-Doyle, K. J., and S. A. Yoon. 2011. "In Search Of Practitioner Based Social Capital: A Social Network Analysis." *Professional Development in Education* 37 (1): 75–93.

Bakhtin, M. 1981. "Discourse in the Novel." In *The Dialogic Imagination*, translated by M. Holquist and C. Emerson, edited by M. Holquist, 259–422. Austin: University of Texas Press.

Bandura, A. 1986. *Social Foundations of Thought and Action: A Social Cognitive Theory*. Upper Saddle River, NJ: Prentice Hall.

BERA (British Education Research Association). 2011. *Ethical Guidelines for Educational Research*. London: BERA.

Bleakley, A. 2006. "You Are Who I Say You Are: The Rhetorical Construction of Identity in the Operating Theatre." *Journal of Workplace Learning*. 17 (7): 414–425.

Bybee, R. W. 2010. "Advancing STEM Education: A 2020 Vision." *Technology and Engineering Teacher* 70 (1): 30–35.

Caza B. B., and S. J. Creary. 2016. "The Construction of Professional Identity." In *Perspectives on Contemporary Professional Work: Challenges and Experiences*, edited by A. Wilkinson, D. Hislop, and C. Coupland, 259–285. Cheltenham: Edward Elgar Publishing.

(CEI) Centre for Education and Industry. 2007. University of Warwick. Careers from Science. *An Investigation for the Science Education Forum*. Warwick: CEI. Accessed August 14, 2017. http://www2.warwick.ac.uk/fac/soc/cei/news/sciencecareersreportv2270607.pdf

Cohen, L., L. Manion, and K. Morrison. 2011. *Research Methods in Education*. 7th ed., 117. Hoboken: Taylor and Francis.

Cooperrider D. L., and S. Srivastva. 1987. "Appreciative Inquiry in Organizational Life." In *Research in Organizational Change and Development*, edited by R. W. Woodman and W. A. Pasmore, Vol. 1, 129–169. Stamford, CT: JAI Press.

Creswell, J. W. 2007. *Qualitative Enquiry and Research Design: Choosing among Five Approaches*. 2nd ed. Thousand Oaks, CA: Sage.

Hargreaves M. 2010. *Educating the Next Generation of Scientists*. HC 492 London: National Audit Office. Accessed December 8, 2017. https://www.nao.org.uk/wp-content/uploads/2010/11/1011492es.pdf

Holman J., and P. Finegold. 2010. *STEM Careers Review. Report to the Gatsby Charitable Foundation*. Accessed December 8, 2017. https://www2.warwick.ac.uk/fac/soc/ier/ngrf/stem/movingon/research/500-stem_careers_review_nov_2010_holman.pdf

Kesidou, S., and M. Koppal. 2004. "Supporting Goals-Based Learning with STEM Outreach." *Journal of STEM Education: Innovations and Research* 5: 5–16.

Koschmann, T. 1999. "Toward a Dialogic Theory of Learning: Bakhtin's Contribution to Understanding Learning in Settings of Collaboration." Article 38 in Proceedings of the 1999 Conference on Computer Supported Collaborative Learning, Stanford University, Stanford, GA, USA.

Laursen, S. L., C. Liston, H. Thiry, and J. Graf. 2007. "What Good is a Scientist in the Classroom? Participant Outcomes and Program Design Features for a Short-Duration Science Outreach Intervention in K-12 Classrooms." *Life Sciences Education* 6 (1): 49–64. doi:10.1187/cbe.06-05-0165.

Laursen, S. L., H. Thiry, and C. Liston. 2012. "The Impact of a University-Based School Science Outreach Program on Graduate Student Participants' Career Paths and Professional Socialisation." *Journal of Higher Education Outreach and Engagement* 16 (2): 47–78.

Lave, J. 1986. "Experiments, Tests, Jobs and Chores: How We Learn to Do What We Do." In *Becoming a Worker*, edited by K. Borman and J. Reisman, 140–155. Norwood, NJ: Albex.

Lave J., and E. Wenger. 1991. *Situated learning: Legitimate Peripheral Participation*. Cambridge, MA: Cambridge University Press.

Maistry, S. M. 2008. "School-University CPD Partnerships : Fertile Ground for Cultivating Teacher Communities of Practice." *South African Journal of Higher Education* 22 (2): 363–374.

Mann, A., and A. Oldknow. 2012. *School-Industry STEM Links in the UK*. London: National Audit Office.

Maychell, K., C. Evans, R. Brooks, B. Lee, and S. Pathak. 1998. *Leaving at 16: A Study of Factors Affecting Young People's Decision to Leave Full-Time Education*. Slough: National Foundation for Educational Research.

Murphy, R., and S. Machin. 2011. *Improving the Impact of Teachers on Pupil Achievement in the UK*. London: The Sutton Trust.

National Audit Office. 2010. *Educating the Next Generation of Scientists*. London: The Stationery Office.

Niemi, P. M. 1997. "Medical Students' Professional Identity: Self-Reflection during the Preclinical Years." *Medical Education* 31 (6): 408–415.

Patton, M. Q. 1990. *Qualitative Evaluation Methods*. 2nd ed. Thousand Oaks, CA: Sage.

Royal Academy of Engineering. 2013. *Enhancing STEM Education in Secondary Schools: Outputs of Engineering Engagement Programme*. Accessed December 8, 2017. http://www.raeng.org.uk/publications/other/book-final-web

Straw, S., and S. Macleod. 2015. *Evaluation of STEMNET's Operations and Impacts 2011–15: Summary Report*. Slough: National Foundation for Education Research.

Turner P. R., K. Fowler, D. Wick, M. Ramsdell, G. Gotham, E. Glasgow, and C. French, eds. 2007. *Math and Science Symposium. BOCES-University Partnership as a Model for Educational Outreach: K-16 STEM Professional Development*. Knoxville, TX.

Vennix, J., P. Brok, and R. Taconis. 2017. "Perceptions of STEM-Based Outreach Learning Activities in Secondary Education." *Learning Environments Research* 20 (1): 21–46.

Vygotsky, L. S. 1978. *Mind in Society: The Development of Higher Psychological Processes*. Cambridge, MA: Harvard University Press.

Walker, B. H., and D. Salt. 2006. *Resilience Thinking: Sustaining Ecosystems and People in a Changing World*. Washington, DC: Island Press.

Wegerif, R. 2011. "Towards a Dialogic Theory of How Children Learn to Think." *Thinking Skills and Creativity* 6 (3): 179–190. doi:10.1016/j.tsc.2011.08.002.

Wenger, E., R. McDermott, and W. Snyder. 2002. *Cultivating Communities of Practice: A Guide to Managing Knowledge*. Boston, MA: Harvard Business School Press.

Yin, R. K. 1984. *Case Study Research: Design and Methods*. London: Sage.

5 The role of non-formal contexts in teacher education for STEM

The case of horno[3] science and technology interactive centre

Claudia Fernández-Limón, Juan Manuel Fernández-Cárdenas and Alma Adrianna Gómez Galindo (Mexico)

ABSTRACT

Teacher education can benefit directly from experiences in non-formal settings. This article presents a research study with elementary teachers who were teaching in public schools in the state of Nuevo León, México, and participated in a STEM Continuous Professional Development (CPD) workshop. The workshop provided a platform for teachers to interact with scientists and disseminators of science, allowing the appropriation of scientific knowledge applied to everyday activities and settings. Participants improved the quality of their teaching practices in classrooms and gained a new understanding of STEM subjects, enabling them to promote inspiring learning experiences with their students, where dialogue, experimentation and elucidation became an important part of their lessons. The study was carried out using ethnographic tools for analysing recorded videos, 15 sets of field notes, and 49 questionnaires. The sequential analysis of talk and gestures in their participation in the CPD workshop demonstrated high levels of involvement, creativity, and collaborative solution of STEM problems.

Background and context

Educational research focusing on teachers' ideas about science in Mexico in the last ten years has been widely documented for teachers in service. Gómez Galindo, García Franco, and García González (2013) reviewed the state of knowledge in this field in Mexico from 2002 to 2011 and identified more than 20 articles on this topic. For instance, Guerra-Ramos, Leach, and Ryder (2005) point out that teachers seem to be able to accommodate their pedagogical ideas into different educational contexts; however, in some cases they also seem to struggle to clarify stereotypical ideas about professional scientific practices. Likewise, the work of García Franco et al. (2006) and García Ruiz and Orozco Sánchez (2008) focuses on the study of attitudes toward science, finding that these are mostly negative and distant from teachers' interests. Regarding disciplinary knowledge, most of the publications refer to teachers at secondary level. Studies carried out by Flores-Camacho et al. (2007), García Franco et al. (2006) and Gallegos, García, and López (2005) found that teachers who participated in Continuous Professional Development (CPD) courses on science do not have sufficient basic knowledge to teach scientific disciplines at secondary level. They also noted that teachers' ideas are similar to those of their students, and share conceptual errors (Calixto Flores 2004). An important conclusion of several of these studies is that the ideas teachers have about science are not directly related to their practice; teachers who took institutional CPD courses on science education and changed some of their ideas about science, did not change their practice in classrooms.

Thus, for teachers in service the possibility of reducing the gap between new curricular proposals, their conceptual misunderstandings, and their performance in school, are related to their opportunity to participate in CPD programmes with the capacity of providing them with tools for personal development, enriching their social conditions and empowering them to overcome the traditionally established educational practices. This is also true in the particular case of science education, where the CPD programmres expect to educate teachers with a contemporary vision and experiences associated with how scientific knowledge is constructed, alongside the ability to design socio-constructivist scenarios for their students' learning progress (Barahona et al. 2014). In addition, these programmes have to consider not only the conceptual challenges for teachers but also their emotional and attitudinal facets. Some studies have shown teachers experiment against a background of feelings of insecurity and vulnerability in their daily work (Sutton and Wheatley 2003), and in teaching science, there is some reluctance to fulfil practical work due to the lack of confidence and sense of being out of control (Guerra-Ramos et al. 2011).

The role of STEM in teacher training

The evidence presented in the previous section demonstrates how traditional scientific teacher education no longer meets the needs of teachers to design and implement innovative educational strategies. We consider that a STEM perspective could be an alternative for traditional scientific education. STEM programmes are characterised by the design of problems where Science, Technology, Engineering and Mathematics must be linked and used in a productive way to find solutions to problems. Due to the diversity of possible problems to be addressed using a STEM perspective, students are exposed to a variety of scientific and engineering activities. The activities may include participation in festivals addressing contemporary issues on energy and sustainability, as well as competitions on robotics, solving rescue challenges, designing solutions for industry and commerce, and even designing the capacity of robots to play football or perform a choreography, to name but a few. Hence, learning is both experiential and situated in the context of solving problems relevant for students and society. This is a productive approach in which we can prepare the next generation of STEM professionals: scientists, engineers, architects and technology professionals to ensure the competitiveness of our society (Sakar 2016). Also, the STEM perspective brings positive experiences for teachers, affording group work and problem solving of situated problems in science.

However, teachers need to be educated to fulfil the demands of these activities, many of them taking place in non-formal scenarios such as museums and after-school clubs. For instance, some of these demands can be expressed in the form of the following questions:

How can students be guided to construct new scientific understandings derived from a STEM experience in which they were involved? How can we help teachers to feel comfortable and competent in these situations? In this article, we will present a case study of teacher involvement in museum activities that will offer some responses to these issues.

STEM culture and hands-on activities: the case of horno³ science museum

Experiential and situated learning are aspects of education where science museums and centres can take a leading role (Falk and Dierking 2000, 2013). Museum horno³ is an interactive centre dedicated to scientific education using the building and historical context of what was a blast furnace for producing steel in the city of Monterrey, Mexico. Most of the activities at horno³ are STEM focused activities. Learners can do field trips and practise Science during 1-day sessions. Activities for children include long courses (10 weekly sessions, 35 hours) in order to develop their interest in pursuing STEM careers. These children's courses have been a tradition since 2007, and follow-up studies indicate that 78% of the participants, who are now more than 18 years old, are studying STEM careers. Nearly 50% of the participants belong to low income families, most of them considered at risk because of violence and poverty, and where school dropout is common (Fernández-Limón 2014). These courses have been taught in different places, including horno³, but also in community centres and private schools. horno³ has developed a partnership with different funders in order to cover the costs of these activities in community centres. Thus, socially responsible companies, the federal government, and the state government of Nuevo León, have sponsored school children, paying for registration fees and transportation to different venues, in public institutions. horno³ has also other outreach activities such as 'Science in the City', which is a free programme where families participate in STEM activities. These activities take place in shopping centres, parks, and other public places. In this respect, the provision of non-formal courses and activities by horno³ seeks to develop participants' problem-solving skills, as close as possible to those used by experts on STEM subjects.

Finally, since 2009, horno³ has implemented a workshop in scientific competencies for teachers. In this workshop, a STEM perspective is implemented transversally to develop scientific conceptual relations, as well as their connection to the world and to contextualised problem solving relevant to communities in the metropolitan area of Monterrey. The proposal of this CPD experience is to contribute to the education of teachers using the affordances of non-formal settings and activities that teachers usually do not have access to in formal CPD programmes. Key practices such as planning before performance, collaborative reviewing processes (Simmons and Lunetta 1993), and the use of contextualised frames of reference towards sustainable social practices (Colucci-Gray et al. 2006), are part of the model used for designing the CPD courses offered for teachers. This model also affords the generation of an award scheme supporting teacher empowerment, and represents an opportunity to explore the legitimacy of museums as productive settings for the development of STEM activities for teacher education. Notably, science museums can provide a framework for involving teachers in environments where teachers themselves are engaged in their own learning tasks.

This study examines the perspective of STEM education situated in non-formal settings, such as the museum horno³, analysing how activities can be seen as an opportunity for dialogic participation and a source of knowledge and confidence for teachers involved in these experiences.

Dialogue and transformative education for STEM activities

Increasingly, transformative learning and dialogue are influencing the development of a pedagogy for teaching STEM subjects (Bell 2016; Clifford and Montgomery 2015; Fernández-Cárdenas 2014; Gutiérrez and Vossoughi 2010; Kazak, Wegerif, and Fujita 2015; Wegerif 2017), and the role of museums and other non-formal settings for educating teachers on STEM subjects has been increasingly recognised. In 1991, the *International Journal of Science Education* published a special issue edited by Lucas on informal sources for learning in science. Also, in 1997 *Science Education* published a special issue on informal science education, co-edited by Dierking and Martin. Lastly, in 2003 Feher and Rennie presented an issue on research in science learning in informal environments in the *Journal of Research in Science Teaching*. In these issues, almost all the papers inquire about the interactions of visitors and exhibitions, focusing on the use of language, learning outcomes and the characteristics of exhibitions. Both Lucas (1991) and Dierking and Martin (1997) pointed out that to reduce the gap between museums and educational research in non-formal settings we needed to broaden the scope in methodological and content domains. In this respect, Feher and Rennie (2003) suggested that one area that has been neglected is the participation of teachers in museums and the various ways in which they can learn in these settings. The role of museums involving teachers so that they value and get enthused about science and technology has not been sufficiently considered.

Although teacher learning in museums has not been sufficiently investigated, this paper aims to help in constructing the research agenda in the area considering the points Rennie and colleagues describe in their work (Rennie et al. 2003). Following Schauble, Leinhardt, and Martin (1997), these authors define three organising themes: (a) learning and learning environments; (b) interpretation, meaning, and explanation; and (c) identity, motivation, and interest. The relation between motivation, museums, and STEM education can derive from the possibility of teachers for attending museums voluntarily, so that their expectations can differ from the alignment with official policies. At the same time, teachers can consider the possibility of using museums as a place to 'play' and have childlike and uninhibited behaviour. The relation between museums, learning, and positive emotional experiences in visitors' responses are well documented (Bell et al. 2009).

Moreover, science museums need to ensure they help wider audiences to get access to the most relevant and updated debates in scientific and technology issues, through participation in exhibits and other non-formal learning activities. Social interaction and communication with peers and experts is crucial for getting access to these debates which usually involve complex issues and problems (Dierking and Falk 1994; Rennie et al. 2003). These interactions help visitors to engage creatively in conversations while manipulating physical and digital artefacts (Salinas, Quintero, and Fernández-Cárdenas 2016). The importance of the analysis of successful cases of learning trajectories in museums is related to the recent demand to establish relationships between formal and non-formal educational settings, and for identifying how people can learn in each environment, the relations of people's everyday knowledge and STEM implications, thus generating successful collaborations and providing an effective science education for all (Stocklmayer, Rennie, and Gilbert 2010).

The incorporation of STEM activities in CPD courses for teachers is not an easy task. Teachers are not used to integrated lessons or to incorporating the engineering thinking process involved in STEM as part of their teaching, because they are usually trained in specific mathematics or science strategies. Engineering training guides participants to follow a process of thinking and the creation of models, whereas in science education, teachers are trained to use classic experimentation as the main pedagogic model. Instead of engineering and scientific training, we need a focus on STEM education addressing simultaneously the development of academic skills applied to the solution of problems with creativity and innovation, as well as promoting the levelling of differences of access to the use of scientific tools of participants from disadvantaged contexts. Thus, we need to help teachers to appropriate the knowledge and skills to achieve these goals, and we believe that one way to get to this point is by promoting dialogic and transformative approaches in educational practices. This is relevant as teachers can learn to support reflections on power relations, which then can be used to make visible the differences between experts and novices, and finding ways to promote dialogic interactions in order to help newcomers to get the required access and knowledge in control of old-timers in a given field of scientific expertise (Lave and Wenger 1991; Wenger 2010). Therefore, participants can be involved in learning trajectories where they can use their shared knowledge of STEM subjects to solve problems in their communities and, in parallel, to meet their social needs.

This is even more crucial, given that formal settings have decreased their influence on the achievement of societal goals in education, due to the lack of capacity for promoting creativity, divergent thinking, and collaboration (Robinson 2010). It has been argued that schooling systems need to transform their industrial mind-set into a more post-industrial millennial era way of working, based on collaboration and innovative ways to construct new institutions capable of facing new urgent problems, such as climate change, economic sustainable development, migration, and the construction of societies which are increasingly plural in terms of their culture and interaction of different lifestyles in cities and suburban areas, thus promoting more action-oriented participatory pedagogies (Dillon 2016). As a result, non-formal settings have become increasingly valued as educational spaces where exploration, creativity, and freedom can relate to the development of skills for ensuring that real needs of communities are taken seriously, so that solutions are jointly constructed. Dewey (1909) has been a solid advocate of the promotion of educational systems where curriculum is directly connected to reality on an everyday basis. In line with this, museums and community centres working on STEM activities have legitimated their mission as valid non-formal settings where academic knowledge becomes socially relevant for social change (see also Bhabha 1994; DiGiacomo and Gutiérrez 2016; Gutiérrez 2008; Rogoff et al. 2016).

Methodology

In this study we use ethnographic tools (Castanheira et al. 2001; Gee and Green 1998; Skukauskaitė, Liu, and Green 2007) in order to study the perspective of participants in their own activities and settings. In educational studies, the aim of using ethnographic tools is to describe the way teachers, students and other types of educational agents are involved in settings such as schools and museums. In a related manner, the analysis of talk in participation in situated activities has been used to understand how participants construct meaning as part of the unfolding sequences of turns taking place in collaborative activities (Goodwin 2006, 2007, 2017). We are using these analyses along with the results of surveys and field notes to triangulate information in order to produce an informed account of the perspective of participants in a CPD course on STEM subjects. In this paper, we are focusing particularly on one session (number 15 in a series of workshops), which is about the construction of hypotheses when using an inclined plane with different materials, angles and types of toy cars. In particular, we look at learning trajectories of teachers in service working in small groups. Teachers in service make predictions about the distance which will be travelled by a toy car when left freely to run from the top of an inclined plane. The aspects they propose looking at include the angle of the plane, the materials of the car, and the initial velocity of the car. One purpose is to establish causal relationships between the aspects involved in the physical phenomena observed in each experiment, and assess to what extent some aspects depend on each other and can be controlled to perform some desired effect and the relation with engineering design.

Participants

This CPD workshop for teachers was offered for the first time at horno[3] in 2009, and was designed by horno[3] and Cinvestav Unidad Monterrey. All the participants have been teachers in service in preschool, primary and secondary levels from public schools.

The presentation of the workshop started on 18 and 19 September 2015, and the last sessions took place on 15 and 16 December 2015. The data presented in this article were collected from session 15, involving 49 participants: 14 were preschool teachers, 17 primary school teachers, and 18 secondary school teachers.

Instruments and procedure

The workshop consisted of 15 three-hour sessions. These sessions were divided in 5 modules:

- Life science
- Matter and energy
- Sustainability
- The teacher as a role model
- New technologies and engineering in the classroom

Session 15, located in the last module, addressed the solution of a problem involving an inclined plane and a toy car. We focused on this session as one example of how participants engaged with high levels of involvement and creativity for the collaborative solution of STEM problems. During this session, the class was arranged to work in small groups of four teachers in each table. Groups seated where they wanted, so they selected their team freely. The material to be used was in the front of the room. The facilitator asked one person of each group to pick up the material in front of the room. When they got back to their teams again, the facilitator gave them the first part of the challenge. They had to build a ramp, so that a car and a truck were able to slide on it. After building the ramp, they had to predict the spot where they think the vehicles would stop. They had to mark the selected spot with a sticker.

The teachers started immediately to build the ramp. The facilitator asked them to first think about the place in the room where they wanted to work, because they were going to try their models. They started playing with the cars simulating ramps without actually building the ramp. The facilitator told them that they could not use the cars and trucks before making their prediction. Only after proposing their hypotheses they were allowed to build the ramp. Some teams added books, notebooks, and chairs to try to elevate the ramps. The conversation focused on height, gravity, the material of the cars and trucks, and the differences in weight for the vehicles. Some teams talked about acceleration, the angle of the ramp related to the distance the car and truck were going to travel. Some teams worked on the tables, some others decided to work on the floor.

When they finished building the ramp, the facilitator told them to put one sticker where they thought the car would stop, and another sticker where they thought the truck would stop. After that, they cut two pieces of different colour yarn covering the same distance predicted for each vehicle and stuck them with masking tape on a given wall. Each team was assigned with a space on a specific wall in the room. In another sticker they had to write: 'This is our prediction for our car', and 'This is our prediction for our truck'. After making their predictions and sticking the yarn on the walls, they had to try the car and the truck and make the measure of the actual distance travelled with a yarn of another colour and stick the yarn next to the other lines with the predictions. The teams talked about weight, distance, gravity, friction, design of the cars and trucks, and stability. All the members of the group were actively participating in the activity.

Results

Analysis of data

An ethnographic case was constructed detailing the trajectory of participants, describing particularly session 15 of the CPD workshop. The case included the following analytical elements:

(a) Categorisation of topics presented by teachers participating in the workshop, using a semantic lexical analysis (Hatch and Brown 1995; Reyes Angona and Fernández-Cárdenas 2015; Reyes Angona, Fernández-Cárdenas, and Martínez Martínez 2013) of the questionnaires administered at the end of the session.
(b) Analysis of talk in interaction (Goodwin 2006, 2017; Goodwin and Heritage 1990) of a transcript of a video with the activities of a small group.
(c) Triangulation of data sources, including detailed ethnographic notes of the session, to increase validity and ensuring we captured the perspective of participants.

Analysis of questionnaires

The answers of the surveyed participants indicate different types of learning. Beyond the appropriation of disciplinary knowledge, the answers revealed the importance that

Table 1. Statements indicating learning with enjoyment.

Statements from questionnaires
'We practice different topics in a different and fun way'
'It made us reflect in a fun way'
'Very entertaining and fun to use in the classroom'
'Very motivating and creative' [regarding activities]

Table 2. Lexical frequencies of terms used for answering the question: What do you think of the session?

Word	Frequency
Applying (and lexical variations such as 'application' and 'applied')	11
Activity (and derived plural forms such as 'activities')	9
Practice (and derived adjectives such as 'practical')	5

participants attributed to the pedagogical model of the course, and the approach focusing on solving problems using a STEM approach. The way of referring to the session allows the identification of three main aspects recognised by participants as relevant in the way they teach STEM subjects:

(a) *Learning with enjoyment:* One of the most challenging obstacles for teachers is the election of a pedagogical model for their practice. They usually struggle to find an efficient pedagogy to disseminate scientific findings and its applications to a wider audience, mostly pupils, and in some cases, parents and policy makers. One of the main challenges of such a pedagogy is the goal of learning with enjoyment, especially in STEM subjects, which in many cases are considered boring and difficult to grasp. The challenge is to develop a pedagogy not only limited to the transmission of useful scientific information, but mainly to draw on a pedagogy aimed to co-construct solutions to STEM problems, while providing enjoyment to the learners. The participants of this CPD course seemed to connect with the same sensations of enjoyment as the goals of the pedagogy required in their own classrooms, in relation to the way in which the course was delivered. Some testimonials are shown in Table 1.

Participants did not indicate that they got distracted from learning objectives nor did they consider that there was any simplification of the value of STEM subjects, despite the enjoyment and entertainment they had as part of the experience.

(b) *Learning by doing:* A semantic analysis of the answers of the survey reveals the lexical density regarding the meaning of 'activity'. The following table of the lexical frequencies in the answers regarding their evaluation of the session shows the prevalence of meanings related to the active and practical knowledge, as well as the application of knowledge. In this analysis, the participants' perspective of the course highlights the importance of learning 'by doing', not only by listening or reading information from a book or transmitted by the instructor (transmitted pedagogical knowledge). Table 2 shows the frequency of the term related to each field, retrieved from the answers of the questionnaires.

To this lexical mapping, we must include the recurrent emergence of phrases that semantically strengthen the feeling of active learning. Some examples are included in Table 3.

(c) *Problem-solving pedagogy:* Participants insisted on the value of the STEM approach for solving problems in their daily environment as being at the centre of the teaching process. Some participants highlighted the relationship between the contents of the session and common topics of interest, both for them and for their students. One of the participants indicated that the session included 'real and interesting topics'. Other statements stand out for the value attributed to the learners, and their participation in the learning experience. Here we present three examples (Table 4).

To sum up, the assessment of the surveyed participants of the session was favourable, especially because they found a positive emotional experience, reasonably interesting, notably active, and most of all, different from what they already had experienced. It is worth mentioning some statements from which this conclusion was obtained: 'Activities that can be applied in an innovative way', 'A different way to approach interesting topics', 'It provides alternatives to raise interest in studying', 'Systematised and planned, but still innovative'. The implication of this analysis for teacher education on STEM subjects resides in the transfer of pedagogic skills and affective experience from their own involvement in this course towards

Table 3. Statements about the visualisation of science as applied knowledge oriented to practical activities.

Example	Comment
'Engineering is a vital tool'	It highlights the condition of engineering as applied science and not as pure hypothetical informative content or knowledge
'due to the experiments we elaborate'	The chosen verb 'elaborate' in first person highlights the active involvement of the subject in such experiments, not only as an observer but also as a constructor of knowledge
'It was experimental, which makes us build our learning'	Awareness of learning as the product of experience, from the action or construction from the subject, and not only as an acquisition or mental reflection

Table 4. Statements related to the perception of a problem-solving pedagogy.

Statement	Comment
'[the session was] very dynamic, allowing people to try it is a good way to study great topics'	Participants experience the session as something that is 'tested', and goes through a comprehensive experience and sensorial subjectivity, not only as an intellectual activity
'The activity was really dynamic and applied a process accordingly to the team'	Flexibility in the establishment of activities. The learners 'arrange' and contextualise the contents
'It allows involving all the students and mobilising their learning'	It highlights the central role and participation of the subject in the activity

the design of activities for the classrooms in which they teach STEM subjects. Participants emphasised the intrinsic potential of some of the activities they experienced to be transferred verbatim to their own classroom. In other cases, it was necessary to contextualise the problem. As reported by one of the participants, they intended to 'apply [the modelled experiment] but in a less complex way for preschool children'. What is predominant, in any case, is the idea of transferring the experiential spirit, and the pedagogical model described above: student-centred, active, problem-solving based, and enjoyable.

Next, in order to exemplify the interaction between teachers and their involvement in the activities, we present the analysis of a transcript of a video recording of the work of a small group using a ramp.

Transcript analysis of a video-recorded session

In this transcript teachers are sliding a toy car through an inclined plane. The toy car slides mainly on a straight line, but eventually skids out of the way towards the left side. The camera turns to the left to focus on another team which has constructed a guarded lane with small cotton balls, so that the car toy does not slide out of its way (Figure 1).

Transcript 1. A car toy following a straight line

Figure 1. Context of the situation.

Figure 2. The wheel is hurt.

Figure 3. Let's see.

F1: Female teacher 1
F2: Female teacher 2
M1: Male teacher 1
M2: Male teacher 2

(1) F1: *The wheel is hurt / get another one / get… (inaudible) little wheel / that is / this wheel does not give* (Figure 2).
(2) F2: *Let's see* (female partner takes the wheel toy) (Figure 3).

Figure 4. What did you put this here for?

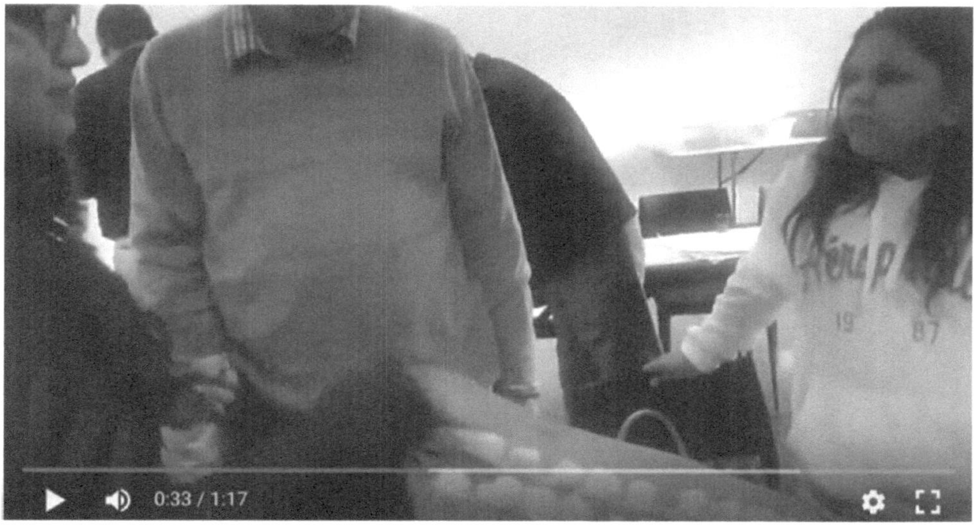

Figure 5. Toy car protection.

(A male participant from another group gets close to F1).

(3) M1: *What did you put this here for?* (Figure 4)

(4) F1: *So that the toy car does not get derailed //* (she moves her hand towards the right, simulating a derailment towards that side) (Figure 5).

(Another male participant hands in a new car which has been tested by him in the back part of the room).

(5) M2: *Look at this* (Figure 6).

Figure 6. Look at this.

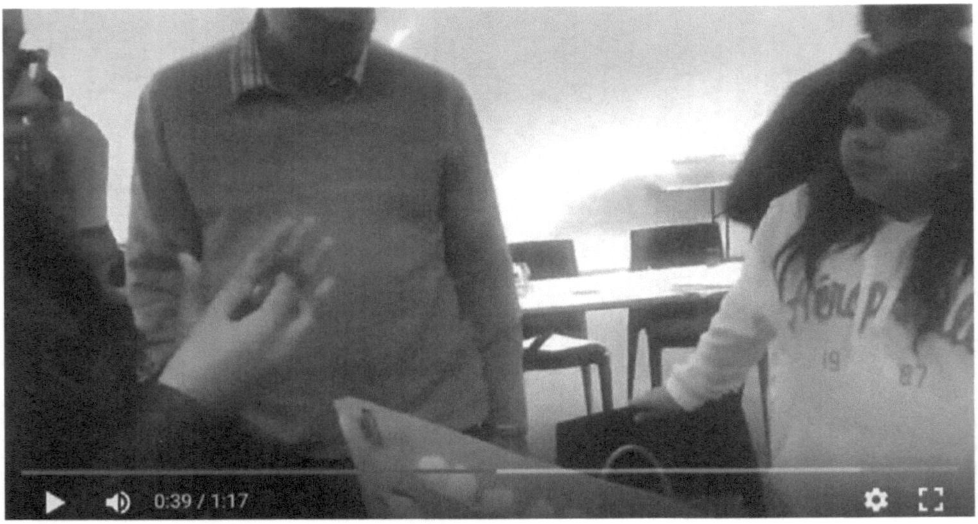

Figure 7. Wheel testing.

(6) F2: *Does it give everything?* (another female participant of the group asks F1 – F1 tries the wheels of the new car sliding them on the palm of her left hand) (Figure 7).

(F1 slides the car toy on the inclined plane, so that the car moves in a straight line without stopping and falling at the end of the table. The camera focus briefly on the activity of the other groups) (Figure 8).

Figure 8. Car sliding.

Analysis of transcript and activity

Teacher participants in the activity are working together in small groups. This specific group is gathered around a long table where they have placed a strip of cardboard as an inclined plane. Within this strip, there is a narrower lane signalled with cotton balls on both sides, so that the toy car can move through a straight line without leaving the lane. One of the female teachers is examining the toy car, testing the wheels as she rubs them on the palm of her hand. While doing this, she mentions that one of the wheels is 'hurt'. She adds that 'this wheel does not give', suggesting that the wheel is stuck, not giving itself in the movement it should follow. It is interesting to note the analogy with a sick body. 'Let's see', a female colleague grabs the toy car. Simultaneously, a male colleague from another team gets close to F1 and asks her on turn 3, 'What did you put this here for?' while pointing to the cotton balls. It is interesting to note the attention these cotton balls get from M1 as the arrangement of participants establishes a clear boundary between groups, so that participants can negotiate together within each group about their own way to design an experiment involving an inclined plane, as well as making their own hypotheses before trying each of their toy cars. Nevertheless, the design of the lane made of cotton balls is apparently attractive enough to get the attention of M1 and get him involved in the business of another team. Heath and vom Lehn (2008) have highlighted the challenges for mutual engagement and co-participation in activities at science museums, yet these teacher participants seem to get as motivated as required to trespass the boundaries of another group and get involved in their design for the experiment using an inclined plane and a toy car.

Both the materiality of the objects (Hetherington and Wegerif, 2018) involved in the experiment and the rules of participation involving an open-ended task seem to exert a strong effect on the expression of creative solutions and interest on the work of others (Fernández-Cárdenas 2008; Fernández-Cárdenas and Silveyra-De La Garza 2010). In addition to this, it is worth noting how simultaneously M2 is looking inside a plastic bag for alternative options of car toys, and testing them in another table sliding them in the surface of such table. He also seems to be crossing boundaries, working in another table, using resources that are not located in the territory of his group.

Eventually, after testing toy car candidates on the table, M2 seems to identify a suitable car for the experiment and grabbing it moves his body through the conversation of F1 and M1, reaching F1 and asking her to look at the car in the palm of his hand. This move incurs in a rupture (Schegloff 1992) of the flow of the sequence of turns between M1 and F1. However, he seems not to care about this and passes the car to F1. F2 is paying attention to this and asks F1 in turn 6 'does it give everything?' Again, it is interesting to note the bodily analogy relating to the car, and the spontaneous involvement of a fourth participant in this unfolding course of cooperative action. Next, F1 lets the car slide freely on the inclined plane. This analogy, and the way teachers interact, shows they are involved in the joint solution of the problem and, at the same time, they use a childlike behaviour playing a game. These results agree with those of Bell et al. (2009) in relation to similar activities encouraging teachers to be creative and spontaneous.

Triangulation: the perspective of participants

From the sources analysed and presented in this article, it is possible to see the high level of engagement of teachers in service as part of this CPD workshop. Participants construct their own trajectories as well as sequentially interacting, making predictions, testing hypothesis, and deciding together about the aspects affecting velocity and trajectories of toy cars and trucks. Teachers reflect on their own involvement creatively so that they propose the transfer of their experience to the design of lessons for their own school classrooms. The data shows how the enthusiasm of participating in an inspirational setting with playful activities is constructive for imagining alternatives for their school students. Data also describes the emergence of a new confidence gained for dealing with STEM subjects in pedagogic situations, and new possibilities for helping their students to interpret what is going in a given attempt for solving a problem using a STEM approach to scientific education. Finally, the data gathered in terms of enthused testimonies and constructive conversations helps to map trajectories which are transformative for the identities of the participants. In other words, they start participating in a more rigid and insecure fashion and end up as enthused participants with a more robust comprehension of the affordances of a STEM approach for solving problems that can be even located in their own community.

Discussion and conclusion

Museums and other non-formal educational settings provide an excellent opportunity for participants to get engaged in scientific activities, appropriating them in a playful manner. In particular, teachers in the analysed session of this CPD workshop were guided in their participation in order to construct scientific knowledge about the use of inclined planes and the impact on velocity of vehicles. We can now revisit the suggestions proposed by Schauble, Leinhardt, and Martin (1997), by defining three organising themes for studying the relationship between museums and scientific education:

(a) Learning and learning environments: Teachers had the opportunity to negotiate the validity of their predictions and get involved in the knowledge of inclined planes as simple machines with everyday uses and situated examples. They also had the opportunity to reflect on the pedagogy that can be better used in their own classrooms.

(b) Interpretation, meaning, and explanation: These pedagogic elements are coupled with an emotionally positive atmosphere in which teachers were feeling free to participate without evaluation outside restrictive environments. Specifically, we consider that this has generated a relationship between collaborative work, a playful environment, and the need to perform an intellectually demanding task in which assertions must be supported by establishing relationships between different aspects of a phenomena and concrete evidence. This combination has afforded relevant learning experiences for teachers for the use of a STEM perspective in their own classrooms.

(c) Identity, motivation, and interest: We suggest that this hands-on experience for constructing scientific knowledge, and all the atmosphere around participants, was more powerful than the usually more theoretical courses teachers have as formal training on science education. The situated and dialogic experience they had at the museum was reported as one of the first opportunities teachers had an authentic involvement in these topics.

Our results help to confirm the relevance of non-formal educational experiences for the construction of knowledge and identities which are usually not afforded by schools and institutions formally designed for teacher education. In this sense, museums in Mexico could play a significant role in CPD courses for educating teachers and in the construction of a STEM culture in schools.

Disclosure statement

No potential conflict of interest was reported by the authors.

ORCID

Claudia Fernández-Limón http://orcid.org/0000-0002-9808-0453
Juan Manuel Fernández-Cárdenas http://orcid.org/0000-0003-2044-1658
Alma Adrianna Gómez Galindo http://orcid.org/0000-0003-3558-0167

References

Barahona, A., J. Chamizo, A. Garritz, and J. Slisko. 2014. "The History and Philosophy of Science and Their Relationship to the Teaching of Science in Mexico." In *International Handbook of Research in History, Philosophy and Science Teaching*, edited by Michael R. Matthews, 2247–2269. New York: Springer.

Bell, D. V. J. 2016. "Twenty-First Century Education: Transformative Education for Sustainability and Responsible Citizenship." *Journal of Teacher Education for Sustainability* 18 (1): 48–56. doi:10.1515/jtes-2016-0004.

Bell, P., B. Lewenstein, A. Shouse, and M. Feder, eds. 2009. *Learning Science in Informal Environments: People, Places and Pursuits*. Washington, DC: The National Academies Press.

Bhabha, H. K. 1994. *The Location of Culture*. London: Psychology Press.

Calixto Flores, R. 2004. "Formas de explicación del proceso de respiración de las plantas." [Forms of Explanation for the Breathing Process of Plants.] *Perfiles Educativos* 26 (105–106): 125–142.

Castanheira, M. L., T. Crawford, C. N. Dixon, and J. L. Green. 2001. "Interactional Ethnography: An Approach to Studying the Social Construction of Literate Practices." *Linguistics and Education* 11 (4): 353–400.

Clifford, V., and C. Montgomery. 2015. "Transformative Learning through Internationalization of the Curriculum in Higher Education." *Journal of Transformative Education* 13 (1): 46–64. doi:10.1177/1541344614560909.

Colucci-Gray, L., E. Camino, G. Barbiero, and D. Gray. 2006. "From Scientific Literacy to Sustainability Literacy: An Ecological Framework for Education." *Science Education* 90 (2): 227–252. doi:10.1002/sce.20109.

Dewey, J. 1909. *Moral Principles in Education/[The Project Gutenberg EBook of Moral Principles in Education]*. Cambridge, MA: Houghton Mifflin. http://www.gutenberg.org/files/25172/25172-h/25172-h.htm.

Dierking, L., and J. H. Falk. 1994. "Family Behavior and Learning in Informal Science Settings: A Review of the Research." *Science Education* 78 (1): 57–72. doi:10.1002/sce.3730780104.

Dierking, L., and L. M. Martin. 1997. "Guest Editorial: Introduction." *Science Education* 81: 629–631.

DiGiacomo, D. K., and K. D. Gutiérrez. 2016. "Relational Equity as a Design Tool within Making and Tinkering Activities." *Mind, Culture, and Activity* 23 (2): 141–153. doi:10.1080/10749039.2015.1058398.

Dillon, J. 2016. "50 Years of JBE: From Science and Environmental Education to Civic Science." *Journal of Biological Education* 50 (2): 120–122.

Falk, J. H., and L. D. Dierking. 2000. *Learning from Museums: Visitor Experiences and the Making of Meaning*. Oxford, UK: Altamira Press.

Falk, J. H., and L. D. Dierking. 2013. *The Museum Experience Revisited*. London: Routledge.

Feher, E., and L. Rennie. 2003. "Guest Editorial." *Journal of Research in Science Teaching* 40 (2): 105–107.

Fernández-Cárdenas, J. M. 2008. "The Situated Aspect of Creativity in Communicative Events: How Do Children Design Web Pages Together?" *Thinking Skills and Creativity* 3 (3): 203–216.

Fernández-Cárdenas, J. M. 2014. "El Dialogismo: Secuencialidad, Posicionamiento, Pluralidad e Historicidad en el Análisis de la Práctica Educativa." [Dialogism: Sequentiality, Positioning, Plurality and Historicity in the Analysis of Educational Practice.] *Sinéctica* 43: 183–203.

Fernández-Cárdenas, J. M., and M. L. Silveyra-De La Garza. 2010. "Disciplinary Knowledge and Gesturing in Communicative Events: A Comparative Study between Lessons Using Interactive Whiteboards and Traditional Whiteboards in Mexican Schools." *Technology, Pedagogy and Education* 19 (2): 173–193.

Fernández-Limón, C. 2014. "2° Estudio de Seguimiento. Diplomados Infantiles." [Second Follow-up Study. Children's certificates.] Educational Assessment 2. Monterrey: horno^3.

Flores-Camacho, F., L. Gallegos-Cázares, X. Bonilla-Pedroza and F. de M. Reyes-Cárdenas. 2008. "Influencia que ejercen la formación y el medio al conformar las concepciones sobre la naturaleza de la ciencia en profesores de ciencias de secundaria." [Influence Exerted by Formation and Environment for Conforming Conceptions About the Nature of Science in Secondary School Teachers.] In *Anuario Educativo Mexicano: Visión Retrospectiva*, edited by G. T. Bertussi and G. González Gómez 179–214. México: Miguel Ángel Porrúa / Universidad Pedagógica Nacional.

Gallegos, L., B. García and L. López. 2005. "Los conceptos básicos de los profesores(as) de biología de secundaria: resultados de una muestra nacional." [The Basic Concepts of Biology in Secondary School Teachers: Results of a National Simple.] In *Memorias Electrónicas del VIII Congreso Nacional de Investigación Educativa*, Hermosillo, México.

García Franco, A., F. Reyes Cárdenas, F. Flores, and L. Gallegos. 2006. "Conocimientos básicos de profesores mexicanos de química de secundaria." [Basic Knowledge of Mexican Secondary School Teachers in Chemistry.] *Educación Química* 17 (3): 379–387.

García Ruiz, Mayra, and Beatriz Sánchez Hernández. 2006. "Las actitudes relacionadas con las ciencias naturales y sus repercusiones en la práctica docente de profesores de primaria." [Attitudes Related to Natural Sciences and their Repercussions on Teaching Practices in Primary School Teachers.] *Perfiles Educativos* 28 (114): 61–89.

García Ruiz, M., and L. Orozco Sánchez. 2008. "Orientando un cambio de actitud hacia las ciencias naturales y su enseñanza en profesores de educación primaria." [Orienting a Change in the Attitude Towards Natural Sciences and their Teaching in Primary School Teachers.] *Revista Electrónica De Investigación Educativa* 7 (3): 539–568.

Gee, J. P., and J. L. Green. 1998. "Discourse Analysis, Learning, and Social Practice: A Methodological Study." *Review of Research in Education* 23: 119–169.

Gómez Galindo, A. A., A. García Franco and C. García González. 2013. "Estado de la Investigación en Educación en Ciencias Naturales, en el Nivel de Educación Básica, durante la década 2002 al 2011."

[State of the Research in Natural Sciences, for the Level of Basic Education During the Decade 2002–2011.] In *Una década de investigación educativa en conocimientos disciplinares en México (2002-2011): matemáticas, ciencias naturales, lenguaje y lenguas extranjeras* [A Decade of Educational Research in Disciplinary Knowledges in Mexico (2002–2011)], edited by Alicia Ávila, Alma Carrasco Altamirano, Alma Adrianna Gómez Galindo, María Guerra Ramos, Guadalupe López Bonilla, and José Ramírez Romero, 165–202. México: COMIE-ANUIES.

Goodwin, C. 2006. "Interactive Footing." In *Reporting Talk: Reported Speech in Interaction*, edited by E. Holt and R. Clift, 16–46. Cambridge: Cambridge University Press.

Goodwin, C. 2007. "Participation, Stance and Affect in the Organization of Activities." *Discourse & Society* 18 (1): 53–73.

Goodwin, C. 2017. *Co-Operative Action*. Cambridge: Cambridge University Press.

Goodwin, C., and J. Heritage. 1990. "Conversation Analysis." *Annual Review of Anthropology* 19: 283–307.

Guerra-Ramos, M., J. García-Horta, D. López-Valentín, A. A. Gómez Galindo, and R. Balderas-Robledo. 2011. "Science Teachers' Attitudes and Perceptions Related to Practical Work: A Self-Report Questionnaire." In *E-Book Proceedings of the ESERA Conference: Science Learning and Citizenship*, Lyon, France, 94–98.

Guerra-Ramos, M., J. Leach, and J. Ryder. 2005. *The Nature of Scientific Enquiry as a Topic of Teaching in Primary Science: Curriculum Expectations, Teachers' Ideas and Implication for Teachers' Education*. Barcelona: ESERA.

Gutiérrez, K. D. 2008. "Developing a Sociocritical Literacy in the Third Space." *Reading Research Quarterly* 43 (2): 148–164.

Gutiérrez, K. D., and S. Vossoughi. 2010. "Lifting off the Ground to Return Anew: Mediated Praxis, Transformative Learning, and Social Design Experiments." *Journal of Teacher Education* 61 (1–2): 100–117. doi:10.1177/0022487109347877.

Hatch, E., and C. Brown. 1995. *Vocabulary, Semantics, and Language Education*. Cambridge: Cambridge University Press.

Heath, C., and D. vom Lehn. 2008. "Configuring 'Interactivity': Enhancing Engagement in Science Centres and Museums." *Social Studies of Science* 38 (1): 63–91.

Hetherington, L., and R. Wegerif. 2018. "Developing a Material-Dialogic Approach to Pedagogy to Guide Science Teacher Education". *Journal of Education for Teaching* 44 (1): 27–43.

Kazak, S., R. Wegerif, and T. Fujita. 2015. "The Importance of Dialogic Processes to Conceptual Development in Mathematics." *Educational Studies in Mathematics* 90 (2): 105–120. doi:10.1007/s10649-015-9618-y.

Lave, J., and E. Wenger. 1991. *Situated Learning: Legitimate Peripheral Participation*. Cambridge: Cambridge University Press.

Lucas, A. M. 1991. "'Info-Tainment' and Informal Sources for Learning Science." *International Journal of Science Education* 13 (5): 495–504.

Rennie, L., E. Feher, L. Dierking, and J. Falk. 2003. "Toward an Agenda for Advancing Research on Science Learning in out-of-School Settings." *Journal of Research in Science Teaching* 40 (2): 112–120.

Reyes Angona, S., and J. M. Fernández-Cárdenas. 2015. "La escritura académica en la formación en línea de docentes de escuelas públicas mexicanas." [Academic Writing in the Online Formation of Teachers in Public Mexican Schools.] *Revista Electrónica de Investigacion Educativa* 17 (2): 1–15.

Reyes Angona S, J. M. Fernández-Cárdenas, and R. Martínez Martínez. 2013. "Comunidades de blogs para la escritura académica en la enseñanza superior. Un caso de innovación educativa en México." [Communities of Blogs for Academic Writing in Higher Education: A Case of Educational Innovation in Mexico.] *Revista Mexica De Investigación Educativa* 18 (57): 507–535.

Robinson, K. 2010. *Changing Education Paradigms*. London: Royal Society for the encouragement of Arts, Manufactures and Commerce. Accessed 21 October 2015. https://www.youtube.com/watch?v=zDZFcDGpL4U&spfreload=10

Rogoff, B., M. Callanan, K. D. Gutiérrez, and F. Erickson. 2016. "The Organization of Informal Learning." *Review of Research in Education* 40 (1): 356–401. doi:10.3102/0091732X16680994.

Sakar, L.. 2016. Harmony STEM Program. C. Fernández, Interviewer, January 29.

Salinas, P., E. Quintero, and J. M. Fernández-Cárdenas. 2016. "Fostering Dialogue in the Calculus Classroom Using Dynamic Digital Technology." *Digital Experiences in Mathematics Education* 2 (1): 21–49. doi:10.1007/s40751-016-0013-9.

Schauble, L., G. Leinhardt, and L. Martin. 1997. "A Framework for Organizing a Cumulative Research Agenda in Informal Learning Contexts." *Journal of Museum Education* 22: 3–8.

Schegloff, E. 1992. "Repair after Next Turn: The Last Structurally Provided Defense of Intersubjectivity in Conversation." *American Journal of Sociology* 1295–1345.

Simmons, P., and V. Lunetta. 1993. "Problem-Solving Behaviors During a Genetics Computer Simulation: Beyond the Expert/Novice Dichotomy." *Journal of Research in Science Teaching* 30: 153–173.

Skukauskaitė, A., Y. Liu, and J. L. Green. 2007. "Editorial Introduction: Logics of Inquiry for the Analysis of Video Artefacts: Researching the Construction of Disciplinary Knowledge in Classrooms". *Pedagogies: An International Journal* 2 (3): 131–137.

Stocklmayer, S., L. Rennie, and J. Gilbert. 2010. "The Roles of the Formal and Informal Sectors in the Provision of Effective Science Education." *Studies in Science Education* 46 (1): 1–44. doi:10.1080/03057260903562284.

Sutton, R., and K. Wheatley. 2003. "Teachers' Emotions and Teaching: A Review of the Literature and Directions for Future Research." *Educational Psychology Review* 15 (4): 327–358.

Wegerif, R. 2017. "Technology and Teaching Thinking: Why a Dialogic Approach is Needed for the Twenty-First Century." In *The Routledge International Handbook of Research on Teaching Thinking*, edited by Rupert Wegerif, Li Li, and James C. Kaufman, 427–440. London: Routledge.

Wenger, E. 2010. "Communities of Practice and Social Learning Systems: The Career of a Concept". In *Social Learning Systems and Communities of Practice*, edited by Chris Blackmore, 179–198. London: Springer. doi: 10.1007/978-1-84996-133-2_11.

6 Public dialogue with science and development for teachers of STEM

Linking public dialogue with pedagogic praxis

Richard Watermeyer and Catherine Montgomery (England)

ABSTRACT

Despite evidence of quality teaching in Science, Technology, Engineering and Mathematics (STEM) subject domains and insistence on the part of many national governments on the economic value of STEM, education, recruitment and retention into STEM subject fields and occupations is said to be continually blighted by a 'leaky pipeline'. In the UK context, schools are seen to benefit from a multitude of external STEM engagement and enrichment providers and initiatives. However, despite evidence of the positive impacts of STEM engagement on learners, there exists a dearth of understanding related to how principles of STEM engagement can facilitate STEM teachers in becoming more pedagogically innovative and relevant and, therefore, engaging of their learners in the classroom context. In this article, we employ a secondary data analysis of two prominent cases of public engagement in science and technology (PEST) in the UK to elicit combined lessons for STEM engagement and the pedagogical development of teachers. We consider the successes of science dialogue in establishing principles of best practice that might be transposed to the development of teachers as more able and effective in the engagement of learners in STEM.

Introduction

Public engagement with science and technology (PEST), the coalescing of non-expert communities and scientists in a process of mutual learning that reveals the cultural relevance of science and science as a cultural practice (AAAS 2016), is championed, most visibly in a policy discourse, for its ameliorative and interconnected effects of enhanced public understanding of, approval for, and trust in science and scientists (Holliman and Jensen 2009). It is also credited for increasing a public enthusiasm and interest in science (Stilgoe, Lock, and Wilsdon 2014). PEST tends in most cases to be implemented on the basis of regulatory concerns of risk and trust (Bucchi 2004; Gauchat 2015) or as an effort to 'upstream' (Wilsdon and Willis 2004) public consensus in the generation of new legislative paradigms for science and technology. Critics, however, signpost the persistence of deficit-informed activities (Irwin 2006); of public engagement 'hitting the notes but missing the mark' (Wynne 2006) in the rehabilitation of public trust in science; and public engagement as a decoy for technocratic governance and publics' manipulation (Cooke and Kothari 2001) where it obfuscates the politics of science and its public participation (Chilvers 2008).

While, the focus of PEST is, habitually undifferentiated, *publics* (a taxonomy which encompasses both adults and children), its relationship with the formal schooling context is weak, tangential or non-existent. Moreover, PEST is not explicitly rationalised on educational terms, even if enhanced scientific literacy is an intrinsic aim. STEM engagement activities conversely, are deliberately targeted towards the engagement of learners, typically those in formal compulsory schooling, in science, technology, engineering and mathematics (STEM) subject disciplines. Furthermore, STEM engagement is distinctive for concentrating on the *early* less latter-stage acculturation of a scientific disposition and identification and nurturing of scientific talent. It is thus more about cultural affirmation than cultural remediation. Yet though PEST and STEM engagement may be driven by discrete foci, these are far from disparate. In fact it may be argued that the one is constitutive or at least facilitative of the other. In this case we suggest that STEM engagement of learners may be transformative and influential to their latter-stage role as scientific citizens, or even as scientists, better accustomed, prepared and motivated to meet the demands of PEST. Notwithstanding, a consensus on STEM engagement remains elusive and obscured by competing interpretations of what STEM constitutes in educational terms (Wong, Dillon, and King 2016). For expediency, we adopt a generalist categorisation of STEM engagement as a smorgasbord of typically extra-curricular activities, events and experiences that are advocated and instigated (in similar terms to PEST) on the basis of locating and cultivating an interest in science among communities that are frequently external to and/or disenfranchised from its conversation.

STEM engagement activities are often seen as synonymous with informal and/or free choice educational spaces like museums and science centres (Falk and Dierking 2000; Stocklmayer, Rennie, and Gilbert 2010; Schwan, Grajal, and Lewalter 2014) and contexts that are manifestly organised around an 'edutainment' logic like science shows (Watermeyer 2013) and science festivals (Jensen and Buckley 2014). They may also have a credential function administered through competition and award schemes (www.crestawards.org) or occur as policy initiatives that seek to kindle the imagination and aspiration of participants of formal schooling through the intervention of STEM role models; such as those provided by science ambassador programmes (Herrmann et al. 2016). A political investment in STEM engagement, specifically, is rationalised in terms of its economic return and a near universal belief by policy-makers that a focus on scientific talent is beneficial to national economic competitiveness (OECD 2010; Royal Society 2010). STEM engagement is also rationalised on the basis of a social mission and as a means towards more equal and fairer participation within and access to formal education and employment opportunities, respectively (see Archer et al. 2012; Archer, DeWitt, and Willis 2013; Archer, DeWitt, and Osborne 2015).

STEM engagement much like PEST is typically organised as participatory activity that brings together scientists and non-experts or learners in conversation and practice. Frequently learners are provided with an opportunity to think and act like scientists. Their engagement is explicitly experiential and intentionally performative. Through various iterations of role-play, STEM engagement is intended to build self-confidence, self-efficacy and agency among learners; especially learners whose interactions and experience of STEM subject disciplines are more limited and/or marginal and constrained by what they lack in science capital (Archer et al. 2015). It is a type of boundary work (Akkerman and Bakker 2011) that facilitates the self-actualisation of learners and their teachers; the efficacy of their self-concept and their capacity to visualise and project future imaginaries in STEM where the borders that demarcate, separate and even antagonise their social and scientific worlds collapse (Watermeyer 2012, 2013).

STEM engagement may thus be experienced as a both a catalyst for social empowerment, social mobility and the future active citizenship of learners, and a means through which to identify and scaffold the development of the best scientific talent. While the two are not necessarily discrete functions, STEM engagement tends, despite its egalitarian overtures, to privilege the cultivation of the 'best' rather than most diverse learner constituencies. In fact, we would argue that despite multiple claims of STEM engagement as a conduit for widened participation, its audience is habitually narrow and selective and dominated by those already predisposed to STEM, often benefitting from an abundance of existing science capital, and recognised by their (nominating) schools as their most able and talented students and/or envoys (see also Dawson 2014).

The role of the teacher in addressing these challenges of engagement is crucial and in some cases STEM engagement has implied the bypassing of the teacher and engendered an emphasis on more direct engagement between scientist and learner. In the UK, the establishment of the Science Learning Centre network in 2004 brought teacher professional development back to the fore and provided a context for science teachers continued professional development (Bishop and Denleg 2006). Notwithstanding, STEM engagement with school-aged learners is a ubiquitous aspect of policy for the popularisation of science and it shares commonalities and crossover with other aspects of the PEST ecosystem such as citizen-juries and *café scientique* from which it may also learn. Moreover, we suggest that there is great value and potential in teachers learning from the technologies of elicitation (Lezaun and Soneryd 2007) operationalised in PEST contexts to remediate organisational deficiencies in STEM engagement and to help them in the early stage embedding of a scientific imagination and/or predilection among those they teach. Crucially we perceive the value in learning from PEST not only in relation to STEM engagement as the exclusive preserve of external, 'informal' education providers but as it might transfer to teachers in schools and influence their greater capacity for pedagogical innovation and ability to engineer and sustain an inclusive and egalitarian classroom environment. This is not to suggest that many teachers are not already pedagogically expert, innovative and successful in the engagement of learners in STEM. Instead the question here is on how these competencies might be further supported and enhanced through the cross-fertilisation of ideas and more of a whole-community approach to teacher development in STEM.

As we have already noted, increasing engagement with STEM has become a priority in many countries and in both the United States and the UK there is a proliferation of initiatives designed to improve learners' engagement with science both within and beyond the classroom (Gamse, Martinez, and Bozzi 2017). Despite the establishment of the Science Learning Centre networks, teacher development and teacher education for STEM education has received much less attention, and policy guidance for teacher education in the UK remains fragmented; although there are examples of exceptionally good practice (including where subject societies offer teaching programmes). In addition to this, whilst teacher education programmes in the UK and internationally habitually have some focus on subject specific pedagogy, this tends to be centred on approaches to assessment, pedagogic discourse and scientific enquiry and the use of ICT rather than on developing teachers' engagement with knowledge of science and judgments on complex scientific issues (Teo and Ke 2014). The erosion of emphasis on subject specific pedagogy is characteristic of changes to teacher education policy in the UK over the last 25 years. With more emphasis on 'school-based' approaches and a broadening of routes into teaching, subject specific pedagogy has been marginalised. Despite recent initiatives in STEM specific schools such as University Technical Colleges (UTCs) in the UK and also in the United States where there are now in excess of 100 specialist STEM schools across 30 states, the training for specialist teachers of STEM remains fragmented.

In this article, we consider how a concern for engaging learners in STEM classrooms, where issues of subject disinterest, disenchantment and attrition continue to prevail (IET 2008), may be mediated by the co-option of alternative methodologies and architectures of PEST. Specifically, this paper focuses on one of the preeminent models of PEST, the public dialogue (Einsiedel 2008), and how public dialogues (or science dialogues) in emergent and controversial science and technology, provide structures for developing teachers' knowledge, pedagogic innovation and engagement in STEM disciplinary contexts. We elucidate principles for success in the science dialogue and propose that these might be transposed to the context of pedagogical development and innovation. Concurrently, we use these principles of best practice in PEST in order to identify opportunities for an approach to STEM engagement that is not restricted to external provision but re-centred on everyday classroom practice. We contend, therefore, that teachers can be significant actors within a PEST ecosystem as much as scientists and professional science communicators are, and they can become what Rose (1999) calls experts of community and thus positive contributors to the enhancement of teaching in STEM. The potential of the dialogical approach enacted in PEST is then, where transposed to the classroom, that which enables teachers to transcend the organisational prescriptiveness and primacy of their role as deliverers of the curriculum to co-conspirators of the scientific imagination.

Methodology

This article offers a secondary analysis of two existing research projects on public dialogue in science and technology and these are outlined in more detail below. Reviewing secondary data sources is an approach which is becoming more prevalent, partly as a result of the increase in open access to data sets. Secondary data analysis is consequently being seen as its own form of empiricism, involving a systematic approach with rigorous evaluative steps (Johnstone 2014). The rapid increase in researchers (particularly in the social sciences and professional disciplines) engaging in time-consuming primary qualitative studies and the availability of this data has engendered efforts to capitalise on (and draw meaning from) this rich existing data-set (Howell Major and Savin Baden 2011). The increased pressure to demonstrate impact in research has also resulted in a need to bring research 'closer to the practitioners who can use it to improve practice' (ibid., 646) and this has led to the further development of secondary data analysis approaches where information from a range of sources is re-analysed for particular aims and purposes. Secondary data analysis or the integration of qualitative (and also quantitative) evidence is distinctive from traditional research review as it entails the extraction of data from existing published sources and the development of a new analysis (Howell Major and Savin Baden 2011). There are a number of methodological approaches to this, but in the case of this study a qualitative meta-analysis was used, where primary qualitative data was synthesised into a theoretical model and principles drawn out (ibid.).

The evaluation reports of the two selected projects were re-analysed to generate critical reflections and implications for teacher development. From this analysis principles were drawn out in order to construct an architecture of dialogue between public engagement with science and technology and teacher development. This paper provides a discussion of critical reflections upon the two public dialogues that combine the first author's direct experience as a dialogue evaluator and his written reports with the input of the second author as an expert in teacher development. This discussion thus triangulates empirical observation of the dialogue phenomenon from an engagement/evaluative perspective with the secondary analysis undertaken through a pedagogical lens, to determine the transferability of an engagement method and its impact(s) in the development of teachers' pedagogical competencies. Crucially, our findings are thus themselves the product of a dual perspective dialogue.

The paper considers directly two cases of public dialogue in science (set within a wider context of a science and society UK government initiative), sponsored by Sciencewise which between 2012 and 2016 was the UK's national centre for public dialogue organised for the explicit purpose of producing better policy in science and technology issues. Whilst both of these dialogues were set in the UK context, this paper aims to draw out principles which could be applied in other international contexts where there is a perceived need for intensified focus on specialised training for STEM teachers (such as Singapore: Teo and Ke 2014). These dialogues explored: (1) patient and public views on the research approval process in health research contexts; (2) public views on new regulation/legislation for mitochondrial replacement technology. Both dialogues involved workshops that were run by professional facilitators and attended by a randomly generated sample of public participants across a range of UK cities. The dialogue workshops also involved scientific specialists and the independent evaluator (Watermeyer and Bartlett 2014; Watermeyer and Rowe 2013). Each workshop featured a broad cross-section of public participants with various social profiles (age, gender, ethnicity). In both cases, the dialogue workshops resulted in tangible operational advice. The first dialogue focused on research approval processes which resulted in the modification of organisational and management protocols. The second dialogue focused on mitochondrial replacement technology and resulted in new regulatory powers and licencing for research in assisted reproductive technology. The evaluation reports of both dialogues, however, also reveal a major contribution to the mutual learning of public and scientific constituencies, an outcome reported in numerous other science dialogue evaluation reports (Watermeyer and Rowe 2013; Rowe and Watermeyer 2016).

The science dialogue

The science dialogue of our experience and discussion is a deliberative event, where non-expert members of the general public are recruited into what is typically a day-long workshop where they are introduced to and asked to discuss a scientific topic of some ethical, social and usually scientific complexity. The deliberative dialogue approach is most frequently used as method for eliciting stakeholder values and opinions (Van de Kerkhof 2006) and knowledge translation and exchange (Lavis et al. 2009) that informs policy development (Culyer and Lomas 2006). The two dialogues of our discussion featured aspects of citizen jury (Crosby 1995) and consensus conference (Joss and Durant 1995) methodologies. Concurrently they operated as a kind of extended peer-review (Funtowicz and Ravetz 1994) and technology assessment (Rip, Misa, and Schot 1995). As deliberative events, dialogue participants were asked to focus and critically reflect upon their own attitudes, perspectives and (no matter how remote or tangential) experiences as related to the science and/or technology and problematise these through collective discussion and consideration of a range of stimulus materials. Discussion was facilitated by a moderator who provided contextual information and posed questions and problems for the dialogue group to resolve. Each dialogue consisted of up to and around twenty participants whose involvement should be noted as voluntary although financially incentivised with a modest payment. They were organised into smaller sub-groups of five or six that worked to specific questions and subsequently fed back at intervals to the entire group.

In the UK, an explicit dimension within the rationale for public dialogue is a contribution to learning. Moreover its impact has been recognised in the terms of influence on 'the knowledge, attitudes and capacity ... to be involved in public dialogue' and in stimulating 'collaboration, networking, broader participation and co-operation in relation to public engagement in science and technology' (Sciencewise 2013, 6). Public dialogue, therefore, is a means of relativising scientific innovation as a public concern and a means of generating a closer interface (through conversation) across public, policy and scientific communities, primarily through the mobilisation of the (inexpert) public as a credible and cogent voice in the discourse of science governance (Wilsdon and Willis 2004). From the perspective of the general public *qua* scientific novice, it is perhaps a unique opportunity from which to exercise democratic citizenship and a means with which to connect social and scientific worlds while evolving a deeper understanding of science and the complexities of its governance. The contribution of public dialogue to the architecture of formal science learning is therefore profound, where the ultimate goal of science education (in a democratic, science and technology economy) is the same ambition of enhanced scientific literacy. Furthermore the complementarity or homogeneity of such goals (PEST and formal education in science) are confirmed by a shared aspiration of equal, inclusive and diverse participation.

Participants across science dialogues uniformly speak of having learnt 'a lot of new things', particularly in relation to the science or technology under discussion. A majority of participants also speak of having had their views changed, though the majority of these same persons also speak of having little prior knowledge of the subject area. Evaluation findings, certainly those emerging from exit-poll questionnaires, explain the success of dialogues in the terms of participation and specifically of having had 'an opportunity to speak and to be heard'; and 'an opportunity to be involved in collective discussion' (Watermeyer and Bartlett 2014, 19). Participants directly speak of their involvement in the respective dialogue as a 'learning experience' and significantly, as that which has made them 'more likely to get involved in [dialogue] events in the future' (Watermeyer and Bartlett2014, 19).

As a site of (informal and though desired, perhaps incidental) learning, the dialogue workshop is explicitly interactional and participatory and mimics the conditions of experiential models of learning (Kolb 1984). It is also explicitly deliberative, with public participants co-opted as independent judges asked to mediate the ethical and social complexity of scientific innovation through a negotiation of different scenarios and response pathways in an attempt to find collective resolutions and consensus and concordantly better options for policy formation. In such a way the public dialogue strongly resonates with forms of student-centred and problem-based models of teaching and learning, which may of course be perceived as both a good and bad thing by teachers. For instance, whilst enabling a more immersive learning experience, these pedagogical approaches may be criticised and avoided by teachers for lacking the kinds of transmissional efficiency, perceived to be preconditional to their satisfying the outcome-oriented and results-driven expectations of what is in the UK, an aggressively neoliberal model of compulsory education.

The role of the dialogue facilitator and classroom teacher are also in parts analogous, with many constructivist models of teaching and learning situating the teacher as not an instructor but facilitator. In such context the manner with which dialogue facilitators mobilise public participants, non-expert in scientific matters and concerns, as competent and confident discussants of science policy, is in ways akin to the teacher stimulating the self-concept and self-actualisation of the classroom learner in STEM. Yet, again we would caution, that the science dialogue and science classroom, as respectively implicit and explicit learning ecosystems, are rationalised on aims that are not necessarily or always congruent. Specifically, new learning is distinguished between the respective contexts for being a desired or essential outcome.

What makes successful dialogue?

Dialogic education emphasises the significance of communicative interactions in learning and privileges the interaction of student voices with the voices of teachers and the curriculum. These dialogic practices allow for elaboration of knowledge in the context of equal social participation (Fernández-Cárdenas 2014) and are seen as particularly crucial in science learning where it is important to consider conversations between paradigms regarding problems and their solutions (ibid.). Despite the espoused importance of dialogic learning, some educationists have voiced concern over the space available for dialogue in the curriculum and the 'authority of scientific knowledge in educational practices' (Fernández-Cárdenas 2014, 3). Here we look to the public dialogue to try to ascertain how this may inform dialogic practices for teachers in schools. There are many social, contextual and educational boundaries to be crossed in this translation of dialogue from the public sphere to the school context but we see teachers, their learning and their engagement with public dialogue as a potential means to translate principles of dialogue from the public to the school setting.

If we are to deduce what accounts for the success of the public dialogue and thus translate it for teachers, we must look towards its social architecture and the manner with which it orchestrates a form of social participation that is explicitly inclusive and welcoming of diversity and which embraces a polyphony of voices (Bakhtin 1984). The aim of this dialogue is to stimulate meaningful conversation within a heterogeneous if not potentially disparate public congregation, whose members hold different if not antagonistic and/or polarised views on a subject, their knowledge of which may be superficial. The starting point of the public dialogue is not then so massively different from the typical baseline of nil that characterises most learners' initial forays in formal science learning. The major difference being, life experience and the greater accumulation of this by public dialogue participants; though this can vary with the public dialogue tending to feature a broad age demographic.

In our evaluations of public dialogue, we have applied the meta-criterion of information translation, which provides the most comprehensive means with which to test the quality and efficacy of the dialogue process (Horlick-Jones, Rowe, and Walls 2007). The translation criterion focuses on the efficiency of information and knowledge transmission and interpretation: comprehensiveness and appropriateness. We have used it to extend and build upon on what Sciencewise (2013) has produced as guidance to high quality dialogue. This guidance stipulates that public dialogue requires clarity, from the outset, in the formalisation of its purpose and objective; that the dialogue occur 'within a culture of openness, transparency and participation'; that it is sufficiently resourced in terms of 'time, skills and funding'; that it adopts a broad approach to question-making/framing that stimulates more holistic discussion; that there is honesty concerning the extent of outcomes and how public input generated through the dialogue will be used (i.e. informing *not* developing policy). Moreover, Sciencewise (2013, 14) advises that dialogue ought to be respectful, equal, non-confrontational, without bias, and accommodating of a plurality of views. This concept of plurality is strong in Bakhtinian dialogue where 'all distance between people is suspended' and there is 'free and familiar contact among people' (Bakhtin 1984, 123; Fernández-Cárdenas 2014).

Adding these aims to the 'translation' criterion, we have sought through our evaluations of public dialogue to determine the extent to which public participants have found the information provided within the respective dialogue to be clear; that they had a clear understanding of why they had been invited to join the dialogue, its aims and the larger process it was a part of, in other words, the *context* of the dialogue and their involvement. Again this has echoes of Bakhtin's construction of dialogue where positioning is key to the construction of successful dialogue and participants' positions (less fixed than roles) change through the processes of social interaction (Bakhtin 1984; Fernández-Cárdenas 2014). Furthermore, in assessing the 'translational efficiency' of the dialogue *process*, we have evaluated (through observation, stakeholder interviews, and exit-poll analysis) the extent to which public participants believe they have been provided an opportunity for full disclosure and that they have said all that they wanted to say; that they have been provided adequate time to discuss all that needed to be covered; and that any summing-up and feedback was accurate and true to their perceptions or accounts of what was discussed.

In one of the best examples we have observed, from the perspective of conducting independent evaluation, the public dialogue on mitochondrial replacement technology (Dialogue 1), a highly ethically contentious and contested area of scientific innovation, highlighted not only what makes for good dialogue but the kinds of transformative learning that can emerge through skilled stewardship (Watermeyer and Rowe 2013, 7–8). These evaluation findings, though our own, point to how a public dialogue process instigates learning from multiple perspectives involving multiple actors, again in the spirit of Bakhtinian plurality, and yet in such ways as illuminates, for non-expert public participants in particular, the ubiquity of science in their day lives. Thus through the example of multi-stakeholder dialogue, comes an opportunity for the teacher to visualise the potential of transforming and scaffolding learners imaginaries in science and the cultivation of learners' self-concept and agency where scientific-exoticism is made visible and accessible within the mundaneness of everyday life (Watermeyer 2013):

> The dialogue has also served as an exercise in active citizenship, with all participants across the workshops, relishing the opportunity to be involved and included in such debate; and as a process of experiential learning with all those involved discovering much about science and

> technology; policy-making processes; about the views of the public (and thereby the diversity and plurality of views held by other participants); and ways with which science and technology are ubiquitous components of public citizens' everyday lives. (Watermeyer and Rowe 2013, 7)

The success in enabling public participants' dialogical interactions with science is also based, as seen from Dialogue 1, on facilitators establishing a safe space for knowledge interactions. Indeed, in this example, the dialogue participants' enjoyment of the dialogue space and a sense of goodwill among participants is credited for the success of knowledge exchange and the confidence and sense of self-efficacy among public participants as 'licenced' contributors to scientific discourse:

> Participants were well-scaffolded throughout the dialogue process; were provided a safe and secure environment with which to openly share their ideas and opinions without fear of appearing foolish, ignorant or open to the reprisal of others. Indeed the dialogue process throughout was characterised by good-will among participants and general sense of bonhomie. The role of the facilitators was integral to this, in generating and maintaining a relaxed, trusting, respectful, good natured and good humoured, and enjoyable dialogue experience among all participants ... in most instances participants reflected that their workshop experience had been highly enjoyable and fulfilling and had bolstered their appreciation not only of science and technology developments (R&D) and science policy issues, but their own everyday interface with these. (ibid.)

There are further Bakhtinian echoes here where 'carnival' is a crucial part of dialogue and 'there is a weakening of ... one-sided rhetorical seriousness' and dogmatism and solemnity are diluted (Bakhtin 1984, 107; Fernández-Cárdenas 2014). The parallels here to the creation of an optimum learning environment are evident and reinforce a need, particularly in the context of formal science education which the research literature tells us is blighted by what can be fierce (habitually gendered) competitiveness among learners, for creating 'collegial' spaces in the cultivation of 'science capital' (Archer et al. 2015). In a similar vein, the success of public dialogue in science, though seemingly self-evident, may be attributed to creating an opportunity for individuals both to hear and be heard and more especially in the context of discussing subject knowledge they may know little of or be disenfranchised from (though it permeates their everyday lives). As such, the contribution of the public dialogue in invoking the agency of the public participant as scientific citizen, is that of the teacher culturing the confidence and capacity of the learner to engage in subject knowledge that may seem out-of-reach:

> Participants in this context clearly relished an opportunity to listen and be heard and to have their say on matters they would ordinarily perceive as out-of-their-grasp yet as the workshops revealed were, in most instances, an integral component of their everyday lives. (Watermeyer and Rowe 2013, 8)

Our final observation of what constitutes good dialogue from this particular case study, is that related to quality moderation and the role of the facilitator in guaranteeing equal opportunity among public participants to hear and be heard:

> The facilitators' multi-faceted role as custodians of fair, inclusive and equitable exchange; as catalysts of dynamic and creative dialogue and in part choreographers of process was in our estimation very well balanced. Facilitators were supportive without suffocating; provided leadership without leading; and appeared true in their interpretation of participants' contributions. Furthermore, we observed genuine skill among the facilitation team when working with difficult and domineering participants. They demonstrated in these instances an ability to cauterize the potential of participants' sabotaging or derailing the dialogue process, yet in such ways which still managed to engage and not segregate more boisterous or troublesome individuals. (ibid.)

In this account, we draw parallels with the role of the teacher in having to monitor, regulate and often police the various 'contributions' of classroom learners and the skill that is required in ensuring parity among learners in their classroom interactions. Here Bakhtin's construct of sequentiality is relevant where speakers begin to appreciate the significance of what precedes his or her intervention and each turn anticipates the future contribution of other interlocutors (Bakhtin 1986; Fernández-Cárdenas 2014). We contrast this with the skill demanded of the teacher (facilitator) in balancing a need to protect the welfare of classroom (dialogue) participants whilst also challenging them to develop further in their thoughts and ideas. This is also consistent with Bakhtin (1984) constructions of polyphony in dialogism where many voices are prioritised over the voice of the teacher. Bakhtin talks about the presence of different voices which together form a worldwide dialogue.

When we delve deeper into the choreography of public dialogue we begin to ascertain a range of other features that might well correspond to the making of superior pedagogical competencies among teachers. We reflect here on both dialogues but particularly the public dialogue on patient and public views of the research approval process (Dialogue 2) as one that produced a significant amount of formative learning in relation to dialogue processes; what is necessary and what is necessary to avoid.

Establishing principles for successful dialogue

The Dialogue 2 evaluation report signposted what we might think of as five core principles to dialogue design and implementation that are applicable to the teaching and learning context. In fact, we would argue that these five principles are as salient to the success of the dialogue facilitator in as much the classroom teacher. The principles focus on the ideal behaviour of facilitators in the dialogue context and their handling of dialogue resources. These principles we suggest are concurrently applicable to the ideal behaviour of teachers in the classroom context and their handling of pedagogical resources.

Principle 1: (Clarity of roles and expectations) calls for an explicit demarcation of dialogue roles, specifically the 'conversational' and 'informational' roles of the facilitator and scientific expert (where present), respectively. Such demarcation will frame the expectations and etiquette of public participants; whom they call upon to help them navigate the 'conversational' and 'informational aspects' of the dialogue, and what they might reasonably expect from such parties.

Principle 2: (Balance and impartiality) is related to Principle 1 and demands that dialogue facilitators and collaborating (scientific) experts insulate the dialogue process from all personal views, beliefs and attitudes towards the dialogue topic and exercise restraint in their contribution to the dialogue, recognising their contribution is *facilitative* and not *generative*. Principle 2 also demands that time be built into the dialogue for verification of dialogue findings and that dialogue participants are allowed to sense-check feedback provided by the dialogue facilitator(s).

Principle 3: (Expansive yet on-point moderation) calls for skilful moderation on the part of the dialogue facilitator and an ability to steer discussion in a way that is exploratory and expansive yet without becoming digressive or meandering.

Principle 4: (Incremental sequencing) calls for appropriate sequencing of the 'informational' and 'conversational' aspects of the dialogue. The dialogue must not necessarily follow a linear design but should adopt an incremental approach to knowledge transmission, assimilation and synthesis; where public participants are provided sufficient content knowledge to sanction their deliberative discussion.

Principle 5: (Appropriate selection and use of stimulus materials and lines of questioning) demands that the dialogue process is supported with appropriate and relevant stimulus materials, chosen with the input of content experts. Concurrently, leading-lines of questioning (that might reflect and reproduce bias) and content-specific or laden questions that might cause to confuse and/or exclude public participants should be avoided.

Transposing principles for successful dialogue to principles for successful teaching

In the context of each of these principles, it is possible to draw parallels with how science teachers, and teachers of any subject discipline for that matter, might approach the task of deepening the engagement of their learners in STEM subject knowledge through dialogue. The principles also, however, where transposed relate to how multi-faceted the teachers' role is and the difficulty of embedding a strategy for successful dialogue into teaching practice. For instance, in principles 1 and 2, where it is argued that the role of the facilitator should be kept distinct from the scientific expert, the teacher's classroom responsibility is conversely both facilitative and generative, conversational and informational. An expectation of the teacher as polymath produces limitations of the kind reflected in dialogue processes, where facilitators may be mistaken for having in-depth subject knowledge when they do not.

In reality, though teachers may have subject expertise, this may be restricted to curriculum content, meaning that the contribution in terms of scientific expertise that exists beyond the school and for instance in universities that supports learner engagement and progression, becomes all the more valuable and important, to a wider, more nuanced and sophisticated understanding of science. Principles 1 and 2 are also significant for what they intimate when transferred to a school context, about reasonable expectations made of teachers and how teachers' pedagogical contributions might be enhanced through external intervention and/or partnering and this in turn broadening not only learners' but teachers' perspectives on content knowledge and mitigating the threat of one-dimensional and/or biased knowledge interpretations and transmissions. Watermeyer and Yeoman (forthcoming) for instance report on how the involvement of university scientists in the school setting provides a valuable interstice to teachers' everyday teaching practice and an opportunity to pause and view their learners and their learning competencies/aptitudes through a different lens, helping them to critically reflect upon, modify and ultimately improve their pedagogical outlook and interactions. This form of dialogue is known to teacher education as it is modelled in the partnerships between university education departments and schools. Teacher educators from a university are often perceived as sources of theoretical knowledge that will support teachers in their practice. This process is recognised as 'professional dialogue' in the teacher development literature (Cheng and So 2012) and is perceived to have a 'positive impact on teachers' knowledge and competence in applying new pedagogies' (ibid., 326).

In the same way, in the public dialogue scientists are provided with fresh perspectives that public participants provide. Take for instance, the following two scientific perspectives on the dialogue process and the gain attributed to the public perspective, further confirmed in another recent evaluation of a dialogue focused on the research priorities of an agricultural research institute (cf. Rowe and Watermeyer 2016)

> What the public said really had merit and was coming at the issues from a different space. I think that if you're less accustomed in these things, you're likely to approach them in a different way and without the baggage of past experience.
>
> It was really interesting to see the participants in the public groups approach and tackle these issues for the first time. I think this whole process provided a fresh pair of eyes. (Watermeyer and Bartlett 2014, 73)

These accounts also articulate a sense of dialogue processes as clearing-spaces that help to mitigate prejudicial or misinformed understandings of science that might be similarly repeated as opportunities in the classroom context for the teacher to recalibrate learners' perspectives, associations and engagement with science. Teachers' own misunderstandings about science are also seen as a barrier to effective engagement with STEM teaching (Watermeyer, Morton, and Collins 2016) and this particular form of dialogue could address areas of teacher misinformation.

With regards to principle 3, it is important to underline that dialogues in science should be both open and expansive yet also disciplined enough so that they do not veer too far from their core focus or become distracted or subsumed by other concerns or interests. In the parallel world of the classroom, the teacher is challenged to both stimulate the imaginative and creative capacities of learners, potentially through participatory and dialogical interaction, and cultivate learners' autonomy and agency, yet without abandoning a focus on the development of subject knowledge and cognate skills. The core message from principle 3 then relates to breadth of knowledge without forgoing depth. The parameters or terms of knowledge acquisition (for dialogical fluency) are re-emphasised in principle 4, which again transposed to the classroom, makes clear how foundational knowledge is prerequisite to the acquisition of learner agency and the mobility and agility of learners in navigating subject knowledge. This principle for successful dialogue chimes loudly with that of successful teaching and learning, which asserts the primacy of time and patience and the incremental development of the learner. Finally, in principle 5, we observe the significance attached to the material contours of dialogue and thus early instruction/knowledge transmission, where the selection of resources designed to ignite and sustain the interest of dialogue participants (learners) is given special focus (see Hetherington and Wegerif, this volume). In this context, we see parallels with museum pedagogy and a partnership between museum 'educators' and classroom teachers in the design, development and implementation of learning materials (cf. Watermeyer 2015)

Discussion

The discipline of teacher education suffers from low status and marginality as a generator of expert knowledge as it has often been compared with longer established or 'traditional' disciplines and seen to be inferior in terms of its knowledge base of professional knowledge and pedagogy (Murray 2012, 598). In this sense, engagement in public science dialogues could raise the status of teachers' knowledge and an engagement with principles of public dialogue as outlined above could provide access to and ways of using that knowledge for teachers and their learners.

Crucially, the science dialogue evidences the plausibility of scaffolding non-expert publics' rapid acquisition and mediation of scientific knowledge, which frequently concerns developments that are controversial and ethically complex and liable to provoke emotive and polarised discussion. Thus we see a significant burden of responsibility placed upon the facilitator. We find also from interrogation of the evaluation accounts, honesty and realism in relation to the conditions and values that serve as enablers for the efficacy of this process, which we argue are transposable to the pedagogical context and to guiding the response of teachers to similar challenges.

What we have established as a set of principles for best practice in science dialogue is in terms of the development of teachers of STEM, transferable, salient and potentially influential to how they might produce and curate conditions for deeper, more immersive, expansive and sophisticated forms of learner engagement and negotiation of (scientifically, socially and ethically) complex knowledge. Moreover these principles, where transposed to the classroom context, provide a values-guide to achieving democratic (and deliberative) participatory learning that: (i) is realistic to and accommodating of potential shortfalls in teachers' subject expertise; (ii) is sympathetic to the restrictions of time and other organisational, cognitive and behavioural determinants such as bias that might undermine or enfeeble the potential of the teacher to affect transformative learning; (iii) centralises the role of the teacher in relativising science and the value of co-opting non-scientific discourses, though risk-laden, in the elucidation of science; and (iv) prioritises through the creation of a safe, inclusive and respectful space, the agency and autonomy of the learner as creative, critical and reflexive thinker. This assertion of values is, we contend, the contribution of the science dialogue to teacher education, in as much as it facilitates delineation of the challenges, needs and potential response protocols for teachers in engaging learners in STEM.

From a process-based perspective, the most successful public dialogues are those that adopt a multi-actor approach and utilise and coalesce expert dialogue facilitators (science communicators or more often than not in the UK, market or quasi social research consultancies); dialogue sponsors (a policy or practitioner community); and scientific experts (those with recognised authority and substantive understanding of the scientific dimensions of the topic being deliberated). However if the quality of public dialogue and by extension publics' scientific literacy depends on the involvement of more than one type of facilitator, what are the implications in transitioning such practice into the classroom and for the development of teacher competencies? We are aware of the challenges presented by the different social, cultural and educational contexts of the public and school spheres. We would argue that teachers require a partnership approach to the development and delivery of curriculum that involves other educational stakeholders be they from the informal educational sector; from universities; or from industry, but that this approach exists largely beyond the current framework for teacher training, which tends to emphasise the competencies of the teacher, typically as a lone-worker.

As one example, Watermeyer (2015) in a study of a museum pedagogy approach to science teaching, reveals both the benefits of teachers working in tandem with museum curators and community outreach workers and an appeal from newly qualified teachers that such an approach forms an integral yet currently absent part of teacher training. Numerous other studies we have been involved in point to the importance of teachers being supported and indeed, having their pedagogical repertoires enriched through exposure to and collaboration with other STEM stakeholders and in the context of STEM careers guidance;

schools-university partnership; and STEM Ambassador schemes. Evaluation studies of these kinds of initiatives tend to focus on the impacts on learners' enthusiasm and interest in science and rather less forcefully elucidate and advocate for the ameliorative effects on teachers. This is an oversight. The long-term impact of any engagement and enrichment intervention depends upon the successful translation, embedding and retention of good practice and what works, which rests fundamentally with the enhanced capacity of teachers to engage their learners.

There are initiatives in both the UK and the United States that aim to promote an integrated approach to the teaching of STEM content areas which enables a shift towards emphasis on 'doing' STEM through authentic inquiry as opposed to simply learning content in discrete disciplinary areas (Sinatra et al. 2017). Research also suggests that an interdisciplinary or integrated curriculum generates 'more relevant, less fragmented, and more stimulating experiences for learners' (Furner and Kumar 2007, 186). Whilst this approach often centres on a particular STEM issue or problem (e.g. Speedometry in Sinatra et al. 2017) it offers a more holistic approach in the same way that public dialogues engender holistic learning from emerging problems of science in society. This integrated (and public dialogic) approach sharpens the question of what skills, beliefs, knowledge bases and experience teachers would need to guide this form of teaching and how training can prepare teachers for this more dialogic form of teaching (Stohlmann, Moore, and Roehrig 2012).

Given that initial teacher education programmes internationally have increasingly taken a 'practicum turn' (McLean Davies et al. 2015) where in-school experience is perceived to have an 'optimal value' in learning to teach (ibid., 514), other forms and sources of knowledge in addition to the traditional teacher education department in the university are being sought. Consequently, the contribution of public dialogue to the development of teacher competencies lies with a recognition that the teacher cannot go it alone and that the success of classroom practice is indelibly linked to a whole-community approach, certainly in the milieu of science.

Disclosure statement

No potential conflict of interest was reported by the authors.

ORCID

Richard Watermeyer http://orcid.org/0000-0002-2365-3771
Catherine Montgomery http://orcid.org/0000-0002-5829-6137

References

AAAS (American Association for the Advancement of Science). 2016. *Logic model for public engagement with science*. Washington, DC: AAAS.

ACME. 2007. *The mathematical needs of 14–19 pathways. Advisory Committee on Mathematics Education and the Royal Society*. London: The Royal Society.

Akkerman, S. F., and A. Bakker. 2011. "Boundary Crossing and Boundary Objects." *Review of Educational Research* 81: 132–169.

Archer, L., E. Dawson, J. De Witt, A. Seakins, and B. Wong. 2015. "'Science Capital': A Conceptual, Methodological, and Empirical Argument for Extending Bourdieusian Notions of Capital Beyond the Arts." *Journal of Research in Science Teaching* 52 (7): 922–948.

Archer, L., J. DeWitt, J. Osborne, J. Dillon, B. Willis, and B. Wong. 2012. "'Balancing Acts": Elementary School Girls' Negotiations of Femininity, Achievement, and Science." *Science Education* 96 (6): 967–989.

Archer, L., J. DeWitt, and J. Osborne. 2015. "Is Science for Us? Black Students' and Parents' Views of Science and Science Careers." *Science Education* 99 (2): 199–237.

Archer, L., J. DeWitt, and R. Willis. 2013. "Adolescent Boys' Science Aspirations: Masculinity, Capital, and Power." *Journal of Research in Science Teaching* 51 (1): 1–30.

Bakhtin, M. M. 1984. *Problems of Doestoevsky's Poetics*. Minneapolis, MN: University of Minnesota Press.

Bakhtin, M. M. 1986. *Speech Genres and Other Late Essays*. Austin: University of Texas Press.

Bishop, K., and P. Denleg. 2006. "Science Learning Centres and Governmental Policy for Continuing Professional Development (CPD) in England." *Journal of In-Service Education* 32 (1): 85–102.

Bucchi, M. 2004. *Science in Society: An Introduction to Social Studies of Science*. Abingdon: Routledge.

Cheng, M. M. H., and W. W. M. So. 2012. "Analysing Teacher Professional Development Through Professional Dialogue: An Investigation into a University-school Partnership Project on Enquiry Learning." *Journal of Education for Teaching*. 38 (3): 323–341.

Chilvers, J. 2008. "Theoretical and Practitioner Perspectives on Effective Participatory Appraisal Practice." *Science, Technology and Human Values* 33 (3): 421–451.

Cooke, B., and U. Kothari, eds. 2001. *Participation: The New Tyranny?*. New York: Zed Books.

Crosby, N. 1995. "Citizens' Juries: One Solution for Difficult Environmental Questions." In *Fairness and Competence in Citizen Participation. Evaluating New Models for Environmental Discourse*, edited by O. Renn, T. Webler, and P. Wiedemann, 157–174. Dordrecht: Kluwer Academic Publishers.

Culyer, A. J., and J. Lomas. 2006. "Deliberative Processes and Evidence-informed Decision-making in Health Care – Do They Work and How Might We Know?" *Evidence & Policy: A Journal of Research, Debate and Practice* 2: 357–371.

Dawson, E. 2014. "'Not Designed for Us': How Science Museums and Science Centers Socially Exclude Low-Income, Minority Ethnic Groups." *Science Education* 98 (6): 981–1008.

Einsiedel, E. F. 2008. "Public Participation and Dialogue." *Routledge Handbook of Public Communication of Science and Technology*, edited by M. Bucchi, and B. Trench, 173–185. Abingdon: Routledge.

Falk, J. H., and L. D. Dierking. 2000. *Learning From Museums: Visitor Experiences and the Making of Meaning*. Walnut Creek, CA: AltaMira Press.

Fernández-Cárdenas, Juan Manuel. 2014. "El Dialogismo: Secuencialidad, Posicionamiento, Pluralidad e Historicidad en el Análisis de la Práctica Educativa." [Dialogism: Sequentiality, Positioning, Plurality and Historicity in the Analysis of Educational Practice] *Sinéctica* 43: 183–203.

Funtowicz, S., and J. R. Ravetz 1994. "The Worth of a Songbird: Ecological Economics as a Post-normal Science." *Ecological Economics* 10 (3): 197–207.

Furner, J. M., and D. D. Kumar. 2007. "The Mathematics and Science Integration Argument: A Stand for Teacher Education." *Eurasia Journal of Mathematics, Science & Technology Education* 3: 185–189.

Gamse, B. C., A. Martinez, and L. Bozzi. 2017. "Calling STEM Experts: How Can Experts Contribute to Students' Increased STEM Engagement?" *International Journal of Science Education, Part B* 7 (1): 31–59.

Gauchat, G. 2015. "The Political Context of Science in the United States: Public Acceptance of Evidence-based Policy and Science Funding." *Social Forces* 94 (2): 723–746.

Herrmann, S., R. M. Adelman, J. E. Bodford, O. Graudejus, M. A. Okun, and V. Kwan. 2016. "The Effects of a Female Role Model on Academic Performance and Persistence of Women in STEM Courses." *Basic and Applied Social Psychology* 38 (5): 258–268.

Holliman, R., and E. Jensen. 2009. "(In)authentic Science and (im)partial Publics: (Re)constructing the Science Outreach and Public Engagement Agenda." In *Investigating Science Communication in the Information Age: Implications for Public Engagement and Popular Media*, edited by R. Holliman, L. Whitelegg, E. Scanlon, S. Smidt, and J. Thomas, 35–52. Oxford: Oxford University Press.

Horlick-Jones, T., G. Rowe, and J. Walls. 2007. "Citizen Engagement Processes as Information Systems: The Role of Knowledge and the Concept of Translation Quality." *Public Understanding of Science* 16 (3): 259–278.

Howell Major, C., and M. Savin Baden. 2011. "Integration of Qualitative Evidence: Towards Construction of Academic Knowledge in Social Science and Professional Fields." *Qualitative Research* 11 (6): 645–663.

IET (Institution of Engineering and Technology). 2008. *Studying Stem: What are the barriers?* Accessed 8 December 17. www.theiet.org/factfiles/education/stem-report-page.cfm?type=pdf

Irwin, A. 2006. "The Politics of Talk: Coming to Terms With the 'New' Scientific Governance." *Social Studies of Science* 36 (2): 299–320.

Jensen, E., and N. Buckley. 2014. "Why People Attend Science Festivals: Interests, Motivations and Self-reported Benefits of Public Engagement With Research." *Public Understanding of Science* 23: 557–573.

Johnstone, M. P. 2014. "Secondary Data Analysis: A Method of which the Time Has Come." *Qualitative and Quantitative Methods in Libraries (QQML)* 3: 619–626.

Joss, S., and J. Durant. 1995. *Public Participation in Science: The Role of Consensus Conferences in Europe*. London: Science Museum.

Kolb, D. A. 1984. *Experiential Learning: Experience as the Source of Learning and Development*. Englewood Cliffs, NJ: Prentice-Hall.

Lavis, J. N., J. A. Boyko, A. D. Oxman, S. Lewin, and A. Fretheim. 2009. "SUPPORT Tools for Evidence-informed Health Policymaking (STP) 14: Organising and Using Policy Dialogues to Support Evidence-informed Policymaking." *Health Research Policy and Systems* 7 (Suppl 1): S14.

Lezaun, J., and L. Soneryd. 2007. "Consulting Citizens: Technologies of Elicitation and the Mobility of Publics." *Public Understanding of Science* 16: 279–297.

McLean Davies, L., B. Dickson, F. Rickards, S. Dinham, J. Conroy, and R. Davis. 2015. "Teaching as a Clinical Profession: Translational Practices in Initial Teacher Education – An International Perspective." *Journal of Education for Teaching* 41 (5): 514–528.

Murray, J. 2012. "Changing Places, Changing Spaces? Towards Understanding Teacher Education Through Space–Time Frameworks." *Journal of Education for Teaching* 38 (5): 597–613.

OECD (Organisation for Economic Co-operation and Development). 2010. *The OECD Innovation Strategy: Getting Head Start on Tomorrow*. Accessed 8 December 17. http://www.oecd.org/dataoecd/3/14/45302349.pdf

Pollard, E., N. Jagger, S. Perryman, M. Van Gent, and K. Mann. 2003. *Ready SET Go: A Review of SET Study and Career Choices*. London: The Institute for Employment Studies and the Engineering and Technology Board.

Rip, A., T. Misa, and J. Schot, eds. 1995. *Managing Technology in Society: The Approach of Constructive Technology Assessment*. London: Thomson.

Rose, N. 1999. *Powers of Freedom: Reframing Political Thought*. Cambridge: Cambridge University Press.

Rowe, G., and R. Watermeyer. 2016. *John Innes Centre engagement on its proposed Science Strategy 2017–2022: Final Report*. London: Sciencewise.

Royal Society. 2010. *The Scientific Century: Securing our Future Prosperity*. London: Royal Society.

Schwan, S., A. Grajal, and D. Lewalter. 2014. "Understanding and Engagement in Places of Science Experience: Science Museums." *Educational Psychologist* 49 (2): 70–85.

Sciencewise. 2013. *The Government's Approach to Public Dialogue on Science and Technology*. Accessed 8 December 17. http://www.sciencewise-erc.org.uk/cms/assets/Publications/Sciencewise-Guiding-PrinciplesEF12-Nov-13.pdf

Sinatra, G. M., A. Mukhopadhyay, T. N. Allbright, J. A. Marsh, and M. S. Polikoff. 2017. "Speedometry: A Vehicle for Promoting Interest and Engagement Through Integrated STEM Instruction." *The Journal of Educational Research* 110 (3): 308–316.

Stilgoe, J., S. J. Lock, and J. Wilsdon. 2014. "Why Should We Promote Public Engagement With Science?" *Public Understanding of Science* 23 (1): 4–15.

Stocklmayer, S., L. Rennie, and J. K. Gilbert. 2010. "The Roles of the Formal and Informal Sectors in the Provision of Effective Science Education." *Studies in Science Education* 46 (1): 1–44.

Stohlmann, M., T. J. Moore, and G. H. Roehrig. 2012. Considerations for Teaching Integrated STEM Education. *Journal of Pre-College Engineering Education Research*: 2 (1): 28–34.

Teo, T. W., and K. J. Ke. 2014. "Challenges in STEM Teaching: Implication for Preservice and Inservice Teacher Education Program." *Theory Into Practice* 53 (1): 18–24.

Van de Kerkhof, M. 2006. "Making a Difference: On the Constraints of Consensus Building and the Relevance of Deliberation in Stakeholder Dialogues." *Policy Sciences* 39 (3): 279–299.

Watermeyer, R. 2012. "A Conceptualisation of the Post-museum as Pedagogical Space." *Journal of Science Communication*. 11 (1): 1–6.

Watermeyer, R. 2013. "The Presentation of Science in Everyday Life: The Science Show." *Cultural Studies of Science Education* 8 (3): 737–751.

Watermeyer, R. 2015. "Science Engagement at the Museum School: Teacher Perspectives on the Contribution of Museum Pedagogy to Science Teaching." *British Educational Research Journal* 41 (5): 886–905.

Watermeyer, R., and A. Bartlett. 2014. *Evaluation of Health Research Authority (HRA) Public and Patient Engagement (PPE) Project*. London: Sciencewise.

Watermeyer, R., and G. Rowe. 2013. *Evaluation of Mitochondria Replacement Consultation*. London: Sciencewise.

Watermeyer, R., and K. Yeoman. Forthcoming. "Issues of Connectivity and Continuity in STEM Engagement: Lessons From a Schools-university Partnership. *International Journal of Science Education, Part B: Communication and Public Engagement*.

Watermeyer, R., P. Morton, and J. Collins. 2016. "Rationalising 'for' and 'against' a Policy for School-led Careers-guidance (SLCG) in STEM in the UK: A Teacher Perspective." *International Journal of Science Education* 38 (9): 1441–1458.

Wilsdon, J., and R. Willis. 2004. *See-through Science: Why Public Engagement Needs to Move Upstream*. Project Report. London: Demos.

Wong, V., J. Dillon, H. King. 2016. "STEM in England: Meanings and Motivations in the Policy Arena." *International Journal of Science Education* 38 (15): 2346–2366.

Wynne, B. 2006. "Public Engagement as a Means of Restoring Public Trust in Science-hitting the Notes, but Missing the Music?" *Community Genetics* 9 (3): 211–220.

7 Innovating science teaching with a transformative learning model

Sandra Gudiño Paredes (Mexico)

ABSTRACT

This exploratory study aimed to describe the impact of the 'Science in Family project', as a transformative learning model for science teachers trying to improve student's attitudes toward STEM subjects. This study took place in a public elementary school in Monterrey, Mexico, which has been developing this project for more than thirteen years with students from 4th, 5th and 6th grade. We used participant observation and interviews with four families whose children are students of this elementary school, and with one family whose sons were students of this school some years ago. Results showed that there is a relationship between positive attitudes towards science in students who were exposed to transformative learning models of teaching. Two of the participants took steps to follow science related careers. This study helps to illuminate the extent to which teacher education models influence students' attitudes and how positive attitudes to science are influenced by the use of learning by doing projects.

Introduction

Traditional practices in science teaching have been questioned by educators who have transformed their practice, moving away from the text book tradition and improving their methodology through reflection (Fraser 2015; Taylor, Taylor, and Luitel 2012). There is also evidence that quality of teaching of science is a high determinant in students' decision about continuing science education after 16 years old (Woolnough 1994). Research on students' attitudes about choosing to study science has been at the centre of the science education research community for more than 40 years (Hendley et al. 1995; Smithers and Robinson 1988). For instance, Ebenezer and Zoller (1993) performed a study with 10th graders about their perception of science topics and found that the most affective variable that influences their attitude towards science was the kind of science teaching that they had experienced. Similarly, a study carried out by Osborne, Simon, and Collins (2003) found six factors that strongly influence student choice for pursuit of a science career; two of these factors centred on the influence that students had from positive experiences of extra class activities and the nature of the activities that occurred within the classroom.

According to Settelmaier (2009) a transformative learning pedagogy looks to enhance students' awareness of the impact that science and technology have on society and also to improve their decision-making skills. Similarly, Bueno (2004) states that this topic should not remain as something mysterious or limited to a formal science laboratory; rather, he recommended the practice of conducting experiments in everyday environments outside the classroom, watching the events or reactions that occur in daily life and using elements and ingredients that could be found at the student's homes. For Espejo and González (2015), transformative learning perceives the human being as immersed in a socialisation process that allows appropriate symbolic models according to the student's own approach.

Transformative learning pedagogy principles are not new, they have been part of the scenery of education for many years. Rousseau (1921) raised the need for returning to nature in order to preserve the human being from a corrupted society, and established the active role of the student in learning so that human knowledge begins with the sensible intuition of things, from which ideas are formed. Mezirow (2003) has extended these ideas to highlight how changes in perspectives can be achieved by collaborating in dialogic practices where different points of view can be contrasted and transformed as part of situated practices.

The Science in Family project reflects what constructivist pedagogues have proposed about involving the child in the learning process through an action based educational design so as to have an impact on the social and political development of their communities (Dewey 1974; Freire 2000). Recently, pedagogy continues to seek more meaningful ways that may enhance students' learning, especially in STEM education. According to a large study on average of students' performance in examinations, active learning as a teaching practice resulted in the highest achievement over the continued use of traditional lecturing system in undergraduate education (Freeman et al. 2014).

Ideas and practice

Context

Monterrey, Nuevo León, is situated in the north of Mexico and has more than four million inhabitants. The elementary school where this case study took place is located in a low-income zone, and has an experienced head teacher in charge with more than 40 years of service. The initiative called 'Science in Family project' emerged from the necessity of having an educational science programme that could, ideally, enhance science curiosity in children from 3rd to 6th grade. The initiative and the whole project was specially designed by the Department of Science and Biology of the Autonomous University of Nuevo León, with the support of the Mexican Science and Technology Council (CONACYT) in order to be performed at home with the participation of all the family members.

Materials given to the students included 6 brochures describing experiments, so that participants can engage in one experiment every two weeks, a notebook as a logbook, colours, pencil and markers. Every fifteen days students and their families have to perform an experiment and collect evidence such as pictures and field notes that would be reviewed by their teacher. There is a contest at the end of the period, and all experiments are exhibited in a science fair and all the notebooks are submitted to a science committee of the state Secretary of Education. The winner of each class, according to the committee, wins a science toy like a microscope or telescope and performs his/her experiment for all in the final science fair (STEMk12dialogue 2016).

The exploratory study involved open interviews and video recording of families of three sixth graders who were performing the project during the fall of 2015 and one family of two male students (19 and 20 years old) who were in this school during their elementary education and had the same teacher that is still in charge of this project. The description of the practice identified these categories which will be examined in turn:

- Family involvement
- Family composition
- Vocational inspiration of students
- Non-formal educational practice

(1) Family involvement:

In almost all the observed families both parents were involved with some variation in the schedule and time dedicated to the experiments; while some of them preferred to work during the weekdays after their regular job, some others worked during the weekends. Usually the father did the hard work like cutting structures or assembling, mothers would take pictures, and the children would write the field notes in their notebook.

(2) Family composition

Family composition observed was mainly traditional, according to Mexican standards, composed by mother, father and offspring. Only one of the families had 3 children, the rest only two. Only one of the mothers was dedicated to home activities during the day, the rest worked outside of the house. None of the parents had a profession related to science, but all of them seemed very enthusiastic about working in the projects with their children.

(3) Vocational inspiration of the students

The response of the students regarding what they would like to study when they grow up was very consistent, mentioning that they would like to become a character related to the science field: scientist, astronaut, science teacher. Only one of the girls wanted to become a lawyer, like her older sister. An important achievement highlighting the influence of this programme on its members was observed in the family with two boys, as both of them were in primary school, participated in this project, and now both of them were on their way to study Biology and Chemistry at university level.

(4) Non-formal educational practice

About this category, we were able to observe that this project enhanced curiosity, not only from children, but also from their entire family, because when they were involved in this kind of practices they seemed to learn and share experiences at the same time. Experiment brochures were used like a guide with steps that sometimes were strictly followed, but at other times participants applied their own creativity adding some other substances or materials, in order to see what would happen. This is how non-formal education took place in these projects.

Discussion and conclusion

In this short paper we have aimed to provide an answer to the question: How can the 'Science in Family project' become a transformative learning model for science teachers that could enhance student's attitudes toward STEM subjects? What we observed in the participating families allows us to infer that a transformative learning approach in teaching science effectively conducts to enhance student's attitudes towards STEM subjects.

Even though some of the participants were still far from becoming higher education students, the majority of them mentioned their desire to choose science as a career. The Science in Family project promotes a transformative learning approach, which conceptualises the human being as immersed in a socialisation process that affords the appropriation of models available in this practice. In this study, it is remarkable how transformative learning occurs also inside the homes of students in a non-formal education setting, where they were learning science by doing. Teachers also play an important role in constructing bridges of interpretation about the phenomena explored at home, so that they take the opportunity to discuss the results of experiments with students at school. More importantly, these activities have a vocational impact on students, helping them to become more confident with a subject that usually is perceived as abstract and complex (Flores-Camacho 2012), re-interpreting it as an activity that can be fun and engaging while crossing boundaries between home and school.

Disclosure statement

No potential conflict of interest was reported by the author.

ORCID

Sandra Gudiño Paredes http://orcid.org/0000-0001-9786-606X

References

Bueno, E. 2004. "Aprendiendo química en casa." [Learning Chemistry at Home.] *Revista Eureka sobre Enseñanza y Divulgación de las Ciencias* 1 (1): 45–51.

Dewey, J. 1974. *On Education*. Chicago, IL: University of Chicago Press.

Ebenezer, J. V., and U. Zoller. 1993. "Grade 10 Students' Perceptions of and Attitudes Toward Science Teaching and School Science." *Journal of Research in Science Teaching* 30 (2): 175–186. doi:10.1002/tea.3660300205.

Espejo, R., and M. González. 2015. "Aprendizaje transformativo y programas de investigación en el desarrollo docente universitario." *REDU. Revista de Docencia Universitaria* 13 (3): 309–330.

Flores-Camacho, F., ed. 2012. *La Enseñanza de La Ciencia En La Educación Básica En México*. México, DF: Instituto Nacional de Evaluación Educativa.

Fraser, S. P. 2015. "Transformative Science Teaching in Higher Education." *Journal of Transformative Education* 13 (2): 140–160. doi:10.1177/1541344615571417.

Freeman, S., S. L. Eddy, M. McDonough, M. K. Smith, N. Okoroafor, H. Jordt, and M. P. Wenderoth. 2014. "Active Learning Increases Student Performance in Science, Engineering, and Mathematics." *Proceedings of the National Academy of Sciences* 111 (23): 8410–8415. doi:10.1073/pnas.1319030111.

Freire, P. 2000. *Pedagogy of the Oppressed*. London: Bloomsbury Publishing.

Hendley, D., J. Parkinson, A. Stables, and H. Tanner. 1995. "Gender Differences in Pupil Attitudes to the National Curriculum Foundation Subjects of English, Mathematics, Science and Technology in Key Stage 3 in South Wales." *Educational Studies* 21: 85–97.

Mezirow, J. 2003. "Transformative Learning as Discourse." *Journal of Transformative Education* 1 (1): 58–63. doi:10.1177/1541344603252172.

Osborne, J., S. Simon, and S. Collins. 2003. "Attitudes Towards Science: A Review of the Literature and its Implications." *International Journal of Science Education* 25 (9): 1049–1079. doi:10.1080/0950069032000032199.

Rousseau, J. J. 1921. *Emile, or Education*. Translated by Barbara Foxley. London: J.M. Dent and Sons.

Settelmaier, E. 2009. *"Adding Zest" to Science Education: Transforming the Culture of Science Classrooms through Ethical Dilemma Story Pedagogy*. Saarbrücken: VDM Publishing.

Smithers, A., and P. Robinson. 1988. *The Growth of Mixed A-Levels*. Manchester, NH: Department of Education, University of Manchester.

STEMk12dialogue. 2016. *Ciencia en Familia*. Accessed November 1, 2017. https://youtu.be/6-379U5tTwc

Taylor, P. C., E. Taylor, and B. C. Luitel. 2012. "Multi-Paradigmatic Transformative Research as/for Teacher Education: An Integral Perspective." In *Second International Handbook of Science Education*, 373–387. Netherlands: International Handbooks of Education. doi: https://doi.org/10.1007/978-1-4020-9041-7_26 .

Woolnough, B. 1994. *Effective Science Teaching*. Buckingham: Open University Press.

8 A hybrid and flipped version of an introductory mathematics course for higher education

N. Patricia Salinas Martínez and Eliud Quintero Rodríguez (Mexico)

ABSTRACT
This in practice paper describes the experience of seven lecturers in a hybrid and flipped version of an introductory mathematics course for higher education. In a Mexican university, lecturers adapted to this innovation supported by an adjusted Massive Open Online Course. The experience revealed the relevance of leaving conventional assessment processes to make way for an understanding of lecturers as a collaborative team, trying to transform their own perspective about the learning of mathematics. This experience is an example of the reconceptualisation of the teaching of STEM education that contributes towards a non-formal educational context, promoting lecturers' education and dialogic transformative learning.

Introduction

The Horizon Report is part of a series of documents published by the global leader New Media Consortium (NMC) presenting on a yearly basis a review of emerging technologies impacting education (Johnson et al. 2016). In the NMC report on Trends for Higher Education it is suggested that a Flipped Classroom could be practised with the integration of Online Learning. Lecturers willing to experience this modality should work in the design of an alternative hybrid course. This model (flipped and hybrid) is based on the idea that outside classrooms students can learn from online videos, thus arriving to class with this task accomplished and promoting a different environment inside the classroom. Teachers can browse courses available online to choose one but, at the same time, are reflecting on their own style for teaching so as to ask questions such as. Is it possible to find videos with our same practice in classroom? Is there a different way to teach? How is digital technology mediating the teaching process? In this paper, we take this reflection as an opportunity for developing a dialogic, transformative learning in lecturers' education.

We work for a university in Mexico that is concerned with innovation in education, including digital technology. Our research agenda in Mathematics Education involves the construction of an alternative pedagogy. Currently, we are using our innovative approach where digital technology integrates to the learning of mathematics (Salinas, Quintero, and Sánchez 2015). Dialogue emerges in the classroom when students interact with the digital technology activities that we designed for the learning of calculus in higher education. The students talk among them and share with the lecturer their thoughts about mathematics, while visualising this concepts with digital technology (Salinas, Quintero, and Fernández-Cárdenas 2016).

Here we want to share briefly the experience of seven lecturers in charge of the introductory mathematics course in our university, who experienced the innovation of the flipped and hybrid model for the introductory mathematics course following our pedagogy. Weekly sessions allowed lecturers to talk about their own experience in their classrooms. The audio recordings of the sessions invite us to identify dialogue as a key element for transforming the practice of lecturers. Dialogue was the mean to resolve and learn from the difficulties encountered with our students.

An adjusted Massive Open Online Course

Massive Open Online Courses (MOOCs), now available across the world, have opened up new avenues for wider audiences in higher education. This is part of the current context of transformation in the educational landscape, which 'offers an opportunity to reconcile all learning spaces by creating synergies between formal education and training institutions and other educational experiences. It also offers new opportunities for experimentation and innovation' (Elfert 2015, 49).

In 2013 Maths and Motion MOOC was produced in Coursera, a leading platform partnering with top universities and organisations in the world to give universal access to Online Learning. Our university connects to that target and we were producing all the video lectures during 2013 for the introductory mathematics course. The pedagogy embedded in the MOOC intends to recall arithmetic, algebra and analytic geometry, in order to use them to understand a problem situation in calculus. We are proposing a change in the conventional teaching of mathematics that usually give students exercises that do not relate to everyday circumstances. Thus, instead of dealing with 'x and y' we include real scenarios such as walking on a straight line, filling up a tank with water, or asking for the temperature changes when climbing a mountain; this gives a situated meaning to 'x and y'. Those scenarios afford opportunities to appreciate mathematical knowledge as a useful tool for understanding the complexity of circumstances and solving situated problems. In our design, the focus on graphic representations afforded by the use of digital technology, invites students to build visually an interpretation of different real situations and triggers problem solving. The involvement of situated meaning and the integration of digital technology stand out as the main feature of this innovation for the teaching and learning of mathematics. In August 2013 Maths and Motion MOOC was released to the world, and set up an opportunity to undertake the challenge of a flipped and hybrid experience for our introductory mathematics students.

A flipped and hybrid experience for the introductory course

Students take the introductory mathematics course when they do not pass (or decide not to take) the mathematics placement test, which could have enabled them to take the first mathematics course of their professional programme. During the Fall semester 2013, students from eight classes experienced the flipped and hybrid version of the same introductory course. In total, we were seven lecturers taking the challenge (one lecturer with 2 classes), each class with 35 students. We were teaching one of the classes and leading the experience. This commitment required great effort from the lecturers throughout the semester, such as attending weekly work meetings where lecturers were sharing their thoughts and doubts about the flipped–hybrid innovation experience.

Before each week, lecturers had several tasks; the first was studying the weekly videos in Maths and Motion MOOC before they met their students in classroom. They had to feel confident with the innovative approach to mathematical knowledge. A second task for lecturers was reviewing the problems that would appear randomly in the weekly test, available to the students through the MOOC platform. As a third task, lecturers should familiarise themselves with the weekly classroom activities that we designed to complement the learning from the video lectures. There were individual and collaborative activities, the majority integrating digital technology. The goal of the activities was to challenge students' learning, seeking deeper understanding, and developing collaborative and digital skills. This required from the lecturers to get familiar with the different software that students used in classroom.

An important design choice of the flipped version was that, at the beginning of each week, students would bring a notebook with a summary of the content learned by studying the videos from the MOOC on Coursera. This represents about 3 to 4 hours to do the homework: studying mathematics through the videos and making summaries in their notebooks. In addition, there were 80-min classroom hours, 4 days a week. Each day students work with the weekly activities: individually, collaboratively, using digital technology, and also taking the weekly test provided by Coursera, having 3 opportunities to improve their grade.

This hybrid and flipped experience supported by the MOOC was an opportunity to foster lecturer's reflection on the transformative learning of mathematics integrating technology. There were 14 weekly sessions performed by the lecturers, all of them audiotaped. The analysis suggests they worked as part of an inclusive space for dialogic relationships. Lecturers were interpreting events from more than two perspectives at once, and their willingness to change their minds was motivated as a response to our university's demand for innovation.

The weekly work meetings, an opportunity for dialogue

Lecturers were aware of the pedagogical innovation we were developing, and the Director of the Mathematics Department invited them to participate with us. The experience required changes in their conventional practices: they should not lecture in classroom or to apply their own exams. Another challenge was supporting students in classrooms to improve their learning using different digital technologies.

We were aiming to offer a suitable situation for lecturers to work together, so that we decided to take advantage of a particular condition of the introductory course: its omission from the formal list of subjects of any academic programme in the university. In this way, the final grade of the introductory course does not have an impact on the students' academic grade average as it is just a course required to be completed. Thus, we proposed lecturers to allow students to self-assess their own performance in the course. The weekly tests from Coursera were a part of the grade, but each daily activity would have also a grade representing the effort each student placed on it. Lecturers agreed to let students self-assess, being guided by a rubric, which is a list of criteria about the expectations for the development of an activity describing levels from excellent to poor quality. At a first glance, lecturers considered this as a risky action, but it was worth taking away from them the responsibility of grading and inviting them to deepen the observation of the behaviour of their students in their classroom. The idea was finally appreciated and a first draft for a rubric was proposed for discussion for each kind of activity, including the video summaries in notebooks and the MOOC' online tests performed in classroom.

The seventh meeting (79 min) was quite singular. We proposed compiling two public lists with 'the good and the bad' for the hybrid and flipped experience and for the conventional practice. Next, we provide a brief empirical evidence of the talk in interaction in that meeting. Lecturer A is the Director of Mathematics Department.

> (Lecturer A, 2:53) Is this 'good' something that has been lost? Something that it was done well before, but today it is not present?
>
> (Leader, 3:01) Yes, ok, or it could be something that was good for myself, I can say, before I already knew the class. I had my notes already done.
>
> (Lecturer A, 3:09) Like I was in a comfort zone.
>
> (Lecturer B, 3:40) I remember I taught this as a 'remedial course' long time ago, but I think that before it was good that we had time for exercising, to practice the algebraic handling... and now I do not see that too much.
>
> (Lecturer A, 4:01) That was done inside the classroom?
>
> (Lecturer B, 4:03) Yes ... and also during homework.
>
> (Lecturer B, 4:14) I see this because now that they are in the Coursera tests I give them a paper sheet and I watch them, and yes, several are doing the algebraic procedures, but I also have students that act like (the lecturer mimics the student speedy gestures) ... I finished! ... But they did not even read! ...

The last quotes are significant because show the symptoms of discomfort from leaving the conventional algebraic practices and resort to the test in Coursera. The following quotations show the need to conceive a different paradigm for evaluation.

> (Lecturer A, 5:37). But, does that happen because before you gave higher supervision to make him do it [the algebraic procedures], and now you are giving him more freedom and leaving him to do it, not depending on you?
>
> (Lecturers B, 5:47) Well ... I don't know why, it could be ... I mean ... it is because the test has 3 opportunities, some students in Friday, you are watching them, and they know that it does not count, because 'I can do it later and now I do it and I check my mistakes so later I do it again and...'
>
> (Leader, 6:31) Ok, well, in Coursera, giving opportunities is a way to improve learning. But this does not necessarily occur; maybe your student does that in order to learn the answer to the exercises that could be present in the next opportunity. But in Coursera they are doing this favouring [the idea] that tests allow learning and that you can do it again and again. You know that we are taking care to have enough exercises in order to have different tests.

The session moved towards sharing different alternatives to deal with students' attitudes. Finally, the decision was made to prove and share different actions that could make lecturers feel at ease with the innovation. These actions included giving a brief template to guide their summary in each video, and making teams to build questions from the videos to ask other teams. Rubrics for each kind of activity were also improved. Students were encouraged to take their 3 opportunities at the Coursera test in classroom, and write procedures in their notebook.

Non-formal mathematics lecturers' education

The atmosphere of the weekly meetings allowed for continuous dialogue with the seven lecturers. Before the experience, lecturers complained about students not doing homework, or copying it just before class, but now they complained about students not doing their video summaries properly. If lecturers complained before about students not paying attention in class, now they complained about students not making the necessary effort to perform the activities in classroom. Being aware that the need to face difficulties is always present, lecturers were motivated to supervise their students' performance and the self-assessment they carried out on a daily basis.

There was a strong finding related to the change of the assessment process: lecturers no longer dealt with student discontent with exam grades because they were aware that it was their own responsibility. To achieve this, it required the lecturers to be constantly reminding students to take this action seriously, and to act honestly on their own commitment to learning. As a result lecturers were satisfied with the different kind of relationship they could now have with each student.

The final report of international comparisons of STEM education highlights the 'problem' of the tendency of schools and lecturers to adhere to a traditional curriculum and an educational approach that is failing to bring students excitement to learn. The situation is reinforced by strong traditions of assessment, maintaining a focus on testing a comprehensive coverage of concepts at a relatively low level of reasoning and problem solving. This report places the solution in a 'dedicated lecturer professional development in conditions that support significant changes in orientation and belief in relation to teaching in mathematics and science' (Marginson et al. 2013, 113).

Maths and Motion MOOC already proposes a change for the teaching and learning of mathematics. The hybrid and flipped experience for the introductory course is a step forward in this direction. The decision to take the risk and abandon lecturer-applied tests to evaluate student performance was an agreement achieved during a dialogic process, allowing lecturers to appreciate their experience and improve their professional development. Lecturers are now prepared to deal with the situations arising in classroom, to better understand student behaviours when facing activities, and acting towards helping students to care for their own learning. Meanwhile, students learned to become capable to cope with the challenge in activities, and to be more responsible of their own learning process. A clear benefit we recognise is the importance of transforming evaluation into an honest self-assessment process, guided by rubrics.

It has been useful to experience the creation of a dialogic space in the weekly meetings, where the seven lecturers could express themselves without the intention to silence each other. It was an inclusive space of dialogue 'within which self and other mutually construct and reconstruct each other' (Wegerif 2008, 353). There, we all talked about our own experiences in classrooms using digital technology, the students' responses, what was good or wrong, how we felt and how to improve it. There was a conversation where, as Fernández-Cárdenas (2014) points out, key aspects of dialogue such as the sequentiality and positioning in turn-taking, and the plurality and historicity of voices took place, allowing us to recognise our own values about mathematics as an academic discipline, and the learning of mathematics. We believe that the innovation in mathematics education demands a dialogical position towards the investigation of educational practices, and for this, digital technology plays an important role.

Disclosure statement

No potential conflict of interest was reported by the authors.

ORCID

N. Patricia Salinas Martínez http://orcid.org/0000-0002-9909-6158
Eliud Quintero Rodríguez http://orcid.org/0000-0003-1904-337X

References

Elfert, M. 2015. *Rethinking Education. Towards a Global Common Good?*. Paris: UNESCO Documents and publications.

Fernández-Cárdenas, J. M. 2014. "El Dialogismo: Secuencialidad, Posicionamiento, Pluralidad e Historicidad en el Análisis de la Práctica Educativa." [Dialogism: Sequentiality, Positioning, Plurality and Historicity in the Analysis of Educational Practice] *Sinéctica* 43: 183–203.

Johnson, L., S. A. Becker, M. Cummins, V. Estrada, A. Freeman, and C. Hall. 2016. *NMC Horizon Report: 2016 Higher Education Edition*. Austin, TX: The New Media Consortium.

Marginson, S., R. Tytler, B. Freeman, and K. Roberts. 2013. *STEM: Country Comparisons: International Comparisons of Science, Technology, Engineering and Mathematics (STEM) Education*. Final Report. Australian Council of Learned Academies.

Salinas, P., E. Quintero, and J. M. Fernández-Cárdenas. 2016. "Fostering Dialogue in the Calculus Classroom Using Dynamic Digital Technology." *Digital Experiences in Mathematics Education* 2 (1): 21–49. doi:10.1007/s40751-016-0013-9.

Salinas, P., E. Quintero, and X. Sánchez. 2015. "Math and Motion: A (Coursera) MOOC to Rethink Math Assessment." In *Learning and Collaboration Technologies: Second International Conference, LCT 2015, Held as Part of HCI International 2015, Los Angeles, CA, USA, August 2–7, 2015, Proceedings*, edited by P. Zaphiris and A. Ioannou, 313–324. Cham: Springer International Publishing.

Wegerif, R. 2008. "Dialogic or Dialectic? The Significance of Ontological Assumptions in Research on Educational Dialogue." *British Educational Research Journal* 34 (3): 347–361. doi:10.1080/01411920701532228.

9 Developing non-formal education competences as a complement of formal education for STEM lecturers

Roy Alonso Terrazas-Marín (Mexico)

ABSTRACT

This paper focuses on a current practice piece on professional development for university lecturers, transformative learning, dialogism and STEM (Science, Technology, Engineering and Mathematics) education. Its main goals are to identify the key characteristics that allow STEM educators to experiment with the usage of non-formal education approaches, and apply this knowledge to generate transformative learning experiences. It also seeks to develop new teaching skills among the participants as they train to create learning environments that challenge students to apply STEM knowledge to solve real life problems. A university in northern Mexico created a bilingual programme that allowed participating lecturers to be trained and take part in an outdoor experience in which they designed and applied challenging undertakings for their students. This in practice piece presents a brief description of a non-formal education workshop, and provides examples on how to conduct non formal educational activities based on formal curricular content. Preliminary results allow us to infer that participating lecturers were motivated to learn new teaching approaches in an active way. Participating scholars noted that they changed the way in which they delivered formal content and this allowed them to experiment with new activities that permitted their students to have a better understanding of STEM subjects and improve their academic performance. They also mentioned that the workshop enabled them to improve their attitude, leadership, social skills and class creativity.

Introduction

Traditionally, university STEM lecturers in northern Mexico, especially the ones who teach engineering subjects, tend to focus on the development of technical knowledge and often forget other important issues such as interpersonal skills, creativity and character development.

This practice piece presents innovative professional development experiences at a northern university in Mexico, where lecturers involved in a continuing professional development course were invited to use a formal educational model based on competences. The competence model is a constructivist approach that focuses on performance. It is related to a student's own context and seeks results and evidence that may serve as proof of the development of objectives, and is linked to the real world. The university prepared a five module course in which they explained the main characteristics of the competences model, its paradigms and the importance of generating learning evidence. Traditional education requires the student to learn information and gain knowledge. The university officials had some experiences on non-formal education, and they realised that it was a good method to help develop competences. Therefore, the university decided to seek a non-formal education specialist to create a lecturing professional development programme that could help educators develop a new mixed (formal/non-formal) approach.

The non-formal education specialist created a course called 'Non-formal education techniques for STEM educators' which included the participation of a psychologist and engineering lecturers of different fields such as mathematics, science and computer sciences. As part of the development of this course, and in order to try a new approach, participating lecturers had to design activities for their students involving an outdoor camp experience in which non-formal education techniques were going to be used along with the instruction of formal content.

The development of non-formal education competences for university lecturers

The course lasted for a total of 30 h across five sessions, of which 20 h were practices *in situ*, and 10 h were homework assignments. As a final requirement, participants had to design several content classes to be taught at the university campus, by using the subjects covered on the course. They also had to prepare for the outdoor experience. This class had to include a STEM subject to be developed at the camp and participating lecturers had to use a scientific approach methodology to create activities that could illustrate the solution of real life problems by the students.

The university asked the STEM bilingual lecturers who took a previous competences course to take the new non-formal education course. The course was entirely practical and it was divided into the following sessions:

Working with non-formal education

This session included the theoretical background of non-formal education based on the influential work of Coombs (1974) and the study case of the Scouts of Mexico, which is the biggest non-formal education association in Mexico. The main objective of this session was to have lecturers engaged by experimenting with non-formal education activities, which they could apply in their classrooms in order to develop values, team building skills, team support and leadership, by using non-traditional techniques such as academic gaming design and ludic activity development.

Structuring non-formal learning situations

The main goal of this session was based on having the lecturers reflect on their teaching style, and design new approaches to learning in which they could experiment with the students. Each teacher had to present at least five teaching activities in which they had incorporated a non-formal education environment. They also had to establish the educational outcomes of each activity and incorporate critical thinking in the design of each learning opportunity. Lecturers were asked to think in terms of complicated issues in their STEM classes and design activities for them.

Developing dialogic communication

One of the main goals of this session was to create activities in which a dialogical approach could be used in practice. In this session, the participating lecturers were divided into two groups: one of those groups discussed controversial topics while the other group analysed their dialogical approach using the seminal work of Bakhtin (1984, 1986; see also Fernández-Cárdenas 2014). A presentation on dialogism was shared with the lecturers, highlighting aspects such as: the contrast between monological and dialogical speech, polyphonic speech, and the characteristics of heteroglossia as a way to promote common knowledge in students.

Both teams discussed controversial issues and analysed their own dialogical approaches in order to present an essay on how they were going to apply dialogism in the design of STEM classroom activities. Furthermore, participating lecturers were asked to include dialogic activities as part of the rubric of their newly designed classes.

Identifying student needs

This subject enabled lecturers to recognise the different student personalities and how to deal with them. Each lecturer had to play the role of a specific student personality and have the other lecturers identify it and deal with their demands. This session also included several activities on how to recognise the different kinds of learning. All participating lecturers had to design different kinds of activities intended to include multiple intelligences according to the model of Gardner (2011) and also the learning modalities proposed by the formative work of Barbe and Swassing (1979).

Creating challenging non-formal STEM activities based on real life problems

This opportunity included the introduction of the learning framework known as 'challenge-based learning', in which lecturers and students have to work with experimental learning, progressive education and critical pedagogy based on the 'Apple Classrooms of Tomorrow-Today' programme (Apple Inc. 2008). This programme represents a collaborative effort with the education community to identify the essential design principles for the twenty-first century by focusing on the relationships that matter most: those between students, lecturers, and curriculum. The participating lecturers had to present one of the previously designed activities and be evaluated by their peers. One key concept here was that lecturers were asked to identify STEM real life problems susceptible to the interest of students. Lecturers were also asked to design activities in which the students could change their role and become instructors and evaluators, so that they could design their own challenges based on their interests.

Student inspiration

Finally, lecturers were instructed on how to motivate students by creating a series of dialogic discussions on teacher experiences and analysing them to identify the key concepts of motivation. Condron (2015) has worked for several years on developing ways to motivate students of any age and the lecturers had to put their work in practice by adding motivation aspects to their new curricular design.

An example of how an activity was designed by a university lecturer

As a matter of exemplification, a brief description of the outdoor activity designed by one of the computer science educators is provided:

> The lecturer wanted students to develop a strong logical and practical background, so he decided to teach the students a mathematical puzzle known as "The Hanoi Towers". He knew that students had to understand how this puzzle worked before he could ask them to code a computer program to solve it. In the past, he had problems teaching this algorithm due to its complexity, and this motivated the teacher to use a non-formal education approach. At the outdoor activity, he explained the rules of the puzzle and designed a contest in which the students had to construct the puzzle game by gathering materials found in the woods. The puzzle consists of three rods, and a number of disks of different sizes which can slide onto any rod. The objective of the puzzle is to move the entire stack to another rod without having a bigger disk over a smaller one. After the students finished this, they played the game for several rounds and took notes of the algorithms as they were experimenting with them. Once the outdoor activity ended, the teacher asked the students to use their newly acquired practical knowledge to code a computer program of the puzzle. The teacher realized that this activity helped the students to understand the concepts better and to enhance dialogical skills as the students discussed how to solve the problem. The workshop allowed the lecturer to include the different features he wanted to improve in the group (aspects such as leadership, social skills, team support, attitude development and values) while the students were solving Science and Mathematical problems, and practising English at the same time.

Discussion

The university required all participating lecturers to attend to all six sessions and submit all the assignments. The university staff also analysed the adjustments of the curricular design in order to give credit to the participating lecturers. All lecturers were required to have at least five non-formal activities designed to explain STEM subject classes and lecturers had to present evidences of the application of the concepts seen on the course. Finally, participating educators were asked to design a class and have the course's staff watch how they developed the application of the contents of the course. The lecturers decided to participate in an outdoor activity as their evaluating session.

Participating educators became motivated because they learned new ways to provide STEM content. They also discussed how to design and deliver better ways to approach knowledge. The participant educators were interviewed and they explained that the fact of designing an outdoor experience motivated them because the students were eager to participate. Some participating educators were afraid to experiment, but after the course they were eager to work in this new way. One of the lecturers stated:

> I never thought students could react in such a positive manner, I experienced how students got together in a new way, I saw how spoiled kids became leaders, they confronted their fears,

developed leadership, showed values, talked about their problems, and found solutions to complex challenges, the most unbelievable part is that they were having fun while they were learning.

Some educators identified non-formal education as a platform that enabled them to prepare better classes, because it ensured data processing in a playful and more participative way by providing knowledge in a multi-intelligence form. They also implied that sharing an outdoor experience with their students created a stronger lecturer-student bond. Furthermore, they mentioned that the students were able to achieve better grades and improve their English skills as a result of their participation in the outdoor activity. The incorporation of a mixed approach (formal/non-formal) seemed to provide a common background that allowed a community of practice (Wenger 1998) to be created. This exercise also incorporated critical reflection and dialogism as part of the problem solving methodology. Ruthven, Hoffman, and Mercer (2011) considered that a dialogical approach includes different points of view, and participating lecturers designed their activities to have this in consideration.

Participating educators designed collective, reciprocal and supportive dialogical communication with their students as part of their new class design. This activity developed a significant sense of their experience, changing the way in which they construct their knowledge and how they deal with the requirement of incorporating changes to their subjects, creating what Mezirow (2003; see also Mezirow and Taylor 2009) conceptualised as a transformative learning experience.

Disclosure statement

No potential conflict of interest was reported by the author.

ORCID

Roy Alonso Terrazas-Marín http://orcid.org/0000-0002-8673-4456

References

Apple Inc. 2008. "Apple classrooms of tomorrow-today: learning in the 21st Century." Accessed October 31, 2017. http://ali.apple.com/ACOT2/global/files/ACOT2_Background.pdf

Bakhtin, M. M. 1984. *Problems of Dostoevsky's Poetics*. Edited and translated by C. Emerson. Minneapolis: University of Minnesota Press.

Bakhtin, M. M. 1986. *Speech Genres and Other Late Essays*. University of Texas Slavic Series 8. Austin: University of Texas Press.

Barbe, W. B., and R. H. Swassing. 1979. *Teaching Through Modality Strengths: Concepts and Practices*. Columbus, OH: Zaner Bloser.

Condron, A. 2015. "How to Motivate Students: Top 12 Ways." TeachHUB. Accessed October 31, 2017. http://www.teachhub.com/top-12-ways-motivate-students.

Coombs, P. H., ed. 1974. *Attacking Rural Poverty: How Nonformal Education Can Help*. Baltimore, MD: The Johns Hopkins University Press.

Fernández-Cárdenas, J. M. 2014. "El Dialogismo: Secuencialidad, Posicionamiento, Pluralidad e Historicidad en el Análisis de la Práctica Educativa." [Dialogism: Sequentiality, Positioning, Plurality and Historicity in the Analysis of Educational Practice] *Sinéctica* 43: 183–203.

Gardner, H. 2011. *Frames of Mind: The Theory of Multiple Intelligences*. New York: Basic Books.

Mezirow, J. 2003. "Transformative Learning as Discourse." *Journal of Transformative Education* 1 (1): 58–63. doi:10.1177/1541344603252172.

Mezirow, J., and E. W. Taylor, eds. 2009. *Transformative Learning in Practice: Insights from Community, Workplace, and Higher Education*. 1st ed. San Francisco, CA: Jossey-Bass.

Ruthven, K., R. Hoffman, and N. Mercer. 2011. "A Dialogic Approach to Plenary Problem Synthesis." In *Proceedings of the 35th Conference of the International Group for the Psychology of Mathematics Education*, edited by B. Ubuz, 4:81–88. Ankara: PME.

Wenger, E. 1998. *Communities of Practice: Learning, Meaning, and Identity*. Cambridge: Cambridge University Press.

Index

Note: Page numbers in **bold** refer to Tables